Ham Radio

Booting Up With Hardware and Radio Attacks

(Unlocking the Power of Ham Radio With Digital Modes)

Patrick Harris

Published By **Kate Sanders**

Patrick Harris

All Rights Reserved

Ham Radio: Booting Up With Hardware and Radio Attacks (Unlocking the Power of Ham Radio With Digital Modes)

ISBN 978-1-7777356-6-1

No part of this guidebook shall be reproduced in any form without permission in writing from the publisher except in the case of brief quotations embodied in critical articles or reviews.

Legal & Disclaimer

The information contained in this book is not designed to replace or take the place of any form of medicine or professional medical advice. The information in this book has been provided for educational & entertainment purposes only.

The information contained in this book has been compiled from sources deemed reliable, and it is accurate to the best of the Author's knowledge; however, the Author cannot guarantee its accuracy and validity and cannot be held liable for any errors or omissions. Changes are periodically made to this book. You must consult your doctor or get professional medical advice before using any of the suggested remedies, techniques, or information in this book.

Upon using the information contained in this book, you agree to hold harmless the Author from and against any damages, costs, and expenses, including any legal fees potentially resulting from the application of any of the information provided by this guide. This disclaimer applies to any damages or injury caused by the use and application, whether directly or indirectly, of any advice or information presented, whether for breach of contract, tort, negligence, personal injury, criminal intent, or under any other cause of action.

You agree to accept all risks of using the information presented inside this book. You need to consult a professional medical practitioner in order to ensure you are both able and healthy enough to participate in this program.

Table Of Contents

Chapter 1: What Is Ham Radio? 1

Chapter 2: Ham Radio Interface 13

Chapter 3: How Ham Radio Works 21

Chapter 4: How To Talk On A Ham Radio 37

Chapter 5: How To Get Or Exercise For A Ham Radio License In 2023 50

Chapter 6: Operational Techniques For A Newly Licensed Ham Radio Operator 73

Chapter 7: Organizing Your Ham Radio Home Station 87

Chapter 8: Setting Up Your Pc With A Ham Radio 97

Chapter 9: Understanding Ham Radio Basics 112

Chapter 10: Setting Up Your First Station 119

Chapter 11: Advanced Operating Techniques 129

Chapter 12: Communicating With Exclusive Hams 148

Chapter 14: Studying For The Ham Radio Technician Exam 162

Chapter 1: What Is Ham Radio?

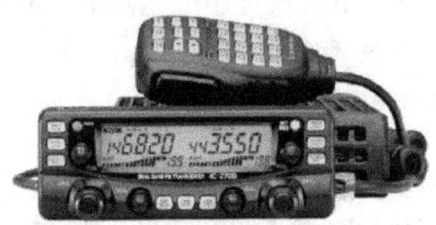

Amateur radio, or HAM radio, is a shape of wireless communiqué that permits people to connect without the want for the net or one-of-a-kind conventional verbal exchange strategies. It is ruled through global governmental corporations and operates on extremely good radio frequencies and with specialized device to broadcast and get maintain of messages.

People from various regions of existence are interested by the hobby of HAM radio, which has extended and fascinating facts. It began all through the flip of the twentieth century on the identical time as wi-fi communiqué have become simply being commenced. The evolution of radio as we understand it today might now not had been feasible without amateur radio operators, who have been frequently at the leading edge of rising era and advances.

There are severa uses for HAM radio in modern society. It is regularly used at the side of formal competitions or sports as a way of conversation amongst HAM radio operators everywhere inside the globe. Others use it to offer emergency communication services in instances of disaster or distinct situations, when conventional conversation traces can be interfered with or unavailable.

A man or woman need to get keep of a license from the applicable government body to lawfully use HAM radio. To do that, the operator ought to pass a take a look at that gauges their technical knowledge, reputation of protection protocols, and know-how of radio guidelines. After receiving their licenses, HAM radio operators can hook up with extremely good operators all around the globe using numerous modes, such as voice, virtual modes, and Morse code.

The device implemented in HAM radio is quite numerous, starting from pocket-sized transportable radios to extra complex base stations with effective transmitters and cutting-edge-day antennas. Many HAM radio operators take satisfaction in developing their private gadget or attempting out out diverse antennas or verbal exchange strategies.

The popularity on independent learning and experimenting is one of the outstanding features of HAM radio. It is suggested for HAM radio operators to educate themselves approximately the workings in their equipment and the physics of radio waves. They are also suggested to test with various antennas, radios, and communication strategies to hone their talents and growth their information.

HAM radio is a treasured beneficial resource in instances of crises and herbal catastrophes.

Emergency employees, governmental corporations, and the general public often get essential verbal exchange offerings from HAM radio operators. When a disaster or emergency movements, HAM radio operators may be the most effective ones with conversation abilities, which makes their education and device crucial.

The thrilling hobby of HAM radio has a wealth of possibilities for education, experimentation, and volunteer art work. There are a ton of net belongings and golf equipment and organizations for your vicinity that might assist you examine extra about HAM radio if you're worried. HAM radio is a satisfying and fascinating interest to pursue, irrespective of your interests in connecting with human beings all through the globe, experimenting with modern-day technology, or supplying emergency communication offerings.

History of Ham Radio

Since its invention in the late 19th century, HAM radio has had an extended and thrilling statistics. Its emergence can be linked to the invention of the telegraph within the middle of the 19th century, which made it possible for human beings to speak over incredible distances through electric powered impulses sent through wires.

People began experimenting with wi-fi verbal exchange as telegraph generation superior, using the in recent times located phenomena of electromagnetic waves to hold messages in the course of the air. Italian scientist and inventor Guglielmo Marconi, who in 1901 sent the first wireless signal over the Atlantic Ocean, changed into one of the forerunners of wi-fi communication.

Since Marconi's fulfillment encouraged others to attempt wi-fi communication, extended-distance verbal exchange among newbie radio fans began to take off. The first newbie radio operators within the United States began broadcasting in the early 1900s, and via the

Twenties, the kingdom had tens of thousands of licensed newbie radio operators.

Ham radio operators had been vital to the increase of commercial radio transmission inside the Nineteen Twenties and Nineteen Thirties. Many of the early radio broadcasters were ham radio enthusiasts who built and ran radio stations the use of their information and device. By experimenting with diverse antennas, radios, and distinct pieces of device, beginner radio operators furthermore made a sizable contribution to the development of radio generation.

Amateur radio operators all over again performed a crucial function in World War II, presenting critical communication offerings to the navy and governmental agencies. Numerous amateur radio operators furnished their offerings as "radio operators" in the armed forces, the usage of their information and gadget to have interaction with brilliant operators in some unspecified time in the future of the globe.

After the conflict, novice radio's enchantment grew even more, and new technology like satellite tv for pc communications and television provided extra opportunities for verbal exchange and experimentation. Amateur radio operators had been essential to the distance application because of the truth they helped astronauts communicate and display satellites and spacecraft. Ham radio is a well-desired hobby in the twenty-first century, with loads of heaps of certified operators worldwide. Technology upgrades have created new opportunities for verbal exchange and experimentation, together with software program application-defined radios and virtual verbal exchange protocols.

During catastrophes and disasters like hurricanes, earthquakes, and wildfires nowadays, newbie radio operators despite the fact that provide vital communique offerings. They additionally use their expertise and equipment to compete in sports like "DXing" and "contesting," which require contacting as many an extended way-off stations as they're able to.

The development of HAM radio is a terrific narrative of exploration, ingenuity, and civic engagement. HAM radio has contributed considerably to the development of cutting-edge communique technology, from its early roots within the telegraph and wireless communication to its use in employer broadcasting, vicinity software, and emergency communications. Several on-line sources and clubs and corporations on your location also can offer assistance and direction in case you're interested in gaining knowledge of extra about HAM radio or getting your license.

Basics of Ham Radio

Ham Radio includes using specialised radio gadget to connect to awesome HAM radio operators at some stage in the globe. The international of radio waves, antennas, and virtual verbal exchange has been spherical for added than a century and is a charming one.

HAM radio is extremely good from considered one of a kind radio conversation strategies like -manner radios and business organisation broadcasting for the motive that it's far a non-income, non-enterprise hobby this is strictly managed through governments everywhere within the globe. To function lawfully, HAM radio operators have to private a license, which they need to get with the beneficial resource of passing an examination that assesses their records of the legal hints, protection precautions, and technical talents had to feature HAM radio system.

Voice, digital modes, and Morse code are only some of the severa communication strategies utilized by HAM radio operators. They can use some element from a number one transportable radio to a modern base station with effective transmitters and current antennas. Governments designate the frequencies used for HAM radio, which can be first-rate from those for company broadcasting or one of a kind radio communique.

When disasters or calamities strike, HAM radio is a essential device. HAM radio operators may also additionally additionally provide critical conversation services to emergency responders, governmental businesses, and the general public if conventional communique routes are broken or unavailable. HAM radio golf equipment and organizations regularly take part in charitable endeavors at the equal time as moreover presenting operators steerage and help.

To hone their abilties and increase their information, HAM radio operators also are advocated to test with severa antenna sorts, radios, and verbal exchange techniques. HAM radio operators also can take part in numerous sports and contests, consisting of DXing, which includes contacting as many a long way-off stations as possible. The interest is promoted via HAM radio golf equipment and groups, which moreover manual experimentation and self-education.

The transnational individual of HAM radio is considered one of its one-of-a-type functions. HAM radio customers may additionally talk with different customers in the course of the globe to alternate thoughts and cultural practices. This encourages international peace and aids in disposing of barriers among severa international places and cultures.

Applications of a Ham Radio

The uses or programs of HAM radio are many, starting from assembly new buddies to offering vital verbal exchange services in instances of crisis. Here are some of the packages of a Ham Radio:

Data Communication

Data communique is one of the most thrilling makes use of for HAM radio. RTTY, packet radio, and digital voice are just a few of the several modes that HAM radio operators may additionally use to transmit and acquire virtual communications. As it lets in the switch of statistics and information that may be vital for reaction operations, this can be a useful device inside the path of crises. In contests and activities in which members compete to look who can create the most connections or trade the most data using virtual modes, HAM radio operators may additionally lease information communication.

DXing/HF Operation

Another nicely-favored HAM radio use is for prolonged-distance communique or DXing. High-frequency (HF) bands are a way of world verbal exchange for HAM radio operators. DXing can be hard since it calls for operators to rent present day gear and strategies to get over hurdles like atmospheric interference. Nevertheless, it is able to be exciting as it allows HAM radio operators to hook up with people from anywhere inside the globe and engage in cultural and informational exchanges.

Chapter 2: Ham Radio Interface

Any HAM radio configuration need to have Ham radio interfaces. They offer a whole lot of abilities which can decorate your communication enjoy and will let you link your radio to a computer or different gadgets. Check out the numerous Ham Radio interfaces beneath:

Icom Interface

This is an interface for HAM radio that is particularly designed to engage with Icom radios. You can use software program program to remotely carry out your radio after connecting it to a computer. The Icom interface usually connects the radio to the laptop through a serial port or USB link.

CAT Interface

CAT, which stands for Computer Aided Transceiver, is a form of HAM radio interface that allows laptop-based radio manage. The radio and laptop are normally related using the CAT interface using a serial port or USB hyperlink. You can use software program software software program to remotely adjust the radio's frequency, mode, and wonderful parameters.

Yaesu Interface

A particular form of HAM radio interface created in particular for Yaesu radios is referred to as a Yaesu interface. You can also use software program application to remotely function your radio after connecting it to a computer. The Yaesu interface normally connects the radio to the pc through a serial port or USB connection.

Kenwood Interface

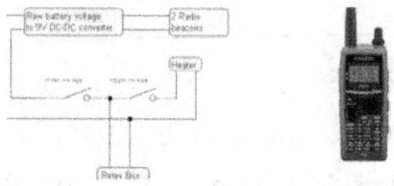

A precise shape of HAM radio interface made for Kenwood radios is known as a Kenwood interface. You can use software program program to remotely characteristic your radio after connecting it to a laptop. The radio and pc are commonly connected thru a serial port or USB hyperlink at the identical time as the usage of the Kenwood interface.

Audio Interface

An audio interface is a specific sort of HAM radio interface that allows you to use an audio connection to hyperlink your radio to a computer or one of a kind devices. It makes it possible to supply and gather audio indicators a number of the pc, the radio, and tremendous devices. RTTY, SSTV, and PSK31 are a few examples of virtual conversation strategies that frequently embody audio interfaces.

PCIe Interface

A unique form of HAM radio interface that attaches to a pc's PCIe slot is called a PCIe interface. Use software program to remotely perform your radio after connecting it to a pc.

Compared to USB or serial port interfaces, PCIe connections often allow higher data transmission speeds, which can also cause faster and additional reliable verbal exchange.

USB Interface

A specific shape of HAM radio interface that attaches to a laptop's USB port is referred to as a USB interface. Use software program application to remotely carry out your radio after connecting it to a computer. The maximum well-known HAM radio interface type is a USB interface, that's frequently easy to set up and use.

FireWire Interface

An interface for HAM radio that hyperlinks to a pc's FireWire connector is referred to as a FireWire interface. Compared to USB or serial port interfaces, FireWire connections are a lot much less commonplace, however they allow faster facts switch prices and may be useful for excessive-bandwidth applications like streaming tune or video.

Any HAM radio gadget need to have HAM radio interfaces. They offer pretty some talents which can enhance your verbal exchange revel in and will let you link your radio to a laptop or one of a kind gadgets. Your specific desires and the sort of radio you very very own will determine the interface kind you choose out.

While PCIe and FireWire connections provide faster records transfer expenses and may be beneficial for excessive-bandwidth applications, USB interfaces are the maximum everyday shape of interface and are regularly easy to install and perform. Regardless of the interface you pick out, make sure you make use of the right software program and cling to the manufacturer's guidelines to assure a reliable and solid verbal exchange enjoy.

What are the troubles with the Audio and CAT interface?

Ham radio customers can speak in speech, virtual, and Morse code, amongst one-of-a-kind methods. For digital mode verbal exchange, radio and computer connections are frequently made through audio and CAT interfaces. Ham radio operators want to be aware that those interfaces may additionally cause troubles.

Here are some of the issues:

1.   Audio stages

Audio stages are some of the most common issues with audio interfaces. It's critical to alter the audio degrees properly at the equal time as connecting a radio to a laptop for digital mode transmission. The distortion and clipping that could cease give up result from immoderate ranges need to make it tough for distinctive operators to decipher the sign. The signal may be too faint to decipher if the tiers are too low. Application can be used to exchange the audio ranges, but it's miles vital to ensure the settings are appropriate for the radio and the program being used.

Solution: Operators need to use an audio interface created expressly for ham radio utilization to save you audio-diploma issues. To make sure that the audio ranges are set well, the ones interfaces often contain built-in stage controls. It's important to stick to the manufacturer's guidelines at the equal time as configuring the interface and converting the audio settings.

2. Ground Loops

Another normal trouble with audio interfaces is floor loops. Multiple floor connections most of the laptop and the radio may additionally reason floor loops, that could introduce undesirable noise into the audio output. Other operators might also discover it hard to decipher the sign due to this noise.

Solution: Use an audio interface with isolation transformers or opto-isolators if you need to save you floor loops. By severing the floor connection among the radio and the pc, those additives useful useful resource within the elimination of ground loops. Additionally, it's important to verify that every one floor connections are set up at a unmarried region, collectively with the power deliver's ground.

three. CAT Interference

Operators may additionally moreover modify the frequency and mode of the radio without touching it with the resource of the usage of CAT interfaces, which can be used to carry out radios from laptop systems. However, CAT connections may also obstruct radio communique, making it difficult for extraordinary operators to pick out out up the signal.

Solution: Operators have to use a CAT interface created expressly for ham radio utilization to save you CAT interference. To lessen interference, those connections frequently comprise filters and isolation circuits. To reduce the hazard of interference, it's also crucial to make certain that the CAT interface is successfully grounded and that the wires are insulated.

Chapter 3: How Ham Radio Works

Amateur radio, or "ham radio," is a provider that lets in radio verbal exchange among human beings. Ham radio operators, often referred to as hams, interact with every other using severa tools and techniques, on the aspect of speech, Morse code, and virtual modes.

Radio Waves

It's important to recognise radio waves to apprehend how Ham radio operates. A shape of electromagnetic radiation known as radio waves bypass thru the environment at the rate of light. They are used for severa sorts of communication, consisting of Wi-Fi, cell mobile phone verbal exchange, radio and television transmission, and similarly.

When an electric powered powered contemporary oscillates to and fro for the duration of an antenna, radio waves are produced. An electromagnetic situation is produced by the oscillating modern and radiates out from the antenna in all commands. Electric and magnetic waves which can be perpendicular to each different make up this area.

Frequency Bands

The frequency bands are among ham radio's maximum important components. Different frequency bands with numerous propagation homes are used for various styles of verbal exchange. Each band within the radio frequency spectrum has various frequencies which might be assigned to it for a specific reason.

In Ham radio, the subsequent frequency stages are most often used:

HF (High Frequency): 3-30 MHz

VHF (Very High Frequency): 30-three hundred MHz

UHF (Ultra High Frequency): 3 hundred-3000 MHz.

Each band has precise propagation homes which could have an impact on how a long way and the way correctly a signal may be acquired. While VHF and UHF signs are regularly restricted to line-of-sight communique, HF indicators, for instance, might also additionally moreover tour big distances by means of manner of bouncing off the ionosphere.

Modes of Communication

In Ham radio, an entire lot of communique channels are used, each having benefits and disadvantages of its very own. The most commonplace modes consist of:

Voice: This is the most normal shape of communique and is corresponding to using the phone. Voice communication can be finished using a headset, a microphone, and a speaker.

Morse Code: A series of dots and dashes is applied in Morse code to symbolize letters and numbers in verbal exchange. A key or pc software software utility can ship Morse code.

Digital Modes: In virtual modes, the sign is encoded and decoded the usage of laptop structures. They can be greater complex modes like JT65 and FT8 or textual content-based absolutely modes like RTTY and PSK31.

Equipment

You need certain device to use Ham radio. The maximum vital tools are as follows:

A transceiver: This is a radio which you use for sign transmission and reception.

An antenna: Signals are sent and obtained thru an antenna. The form of antenna you need relies upon depend upon the frequency band you are the usage of and your place. Antennas come in a massive form of sizes and designs.

A strength source: You can use a battery or a strength supply that plugs into the wall to strength your radio.

Operating Procedures

Operating pointers for Ham radio are in place to make sure that everyone can speak correctly and securely. These steps embody:

Call Signs: Each ham radio operator has a call signal that serves as their radio frequency identification. Government agencies deliver name signs and symptoms which can be used to turn out to be privy to the operator and their vicinity.

Q Codes: Used to unexpectedly transmit specialized information, Q Codes are a series of three-letter codes. As an instance, "QTH" stands for "What is your place?"

Operating Etiquette: On the airwaves, ham radio operators are presupposed to be polite and considerate. This includes the use of suitable jogging techniques, refraining from using vulgar language, and no longer interfering with unique operators.

Regulations

Government prison hints manage Ham radio, and customers are required to abide through the use of way of those pointers.

These laws differ with the aid of u . S . A . And area, however they may be all meant to assure that ham radio users use their device sensibly and appropriately.

The Federal Communications Commission (FCC) within the US is in fee of overseeing novice radio. The FCC offers licenses to operators and establishes recommendations for the usage of radio gadget. To use a Ham radio inside the US, you want to get a license with the useful resource of passing a test.

Along with true law, the ham radio network additionally establishes its very personal suggestions and regulations. There are regulations for the manner to apply numerous conversation techniques, a manner to installation your tool, and how to conduct yourself whilst the use of the radio.

Propagation

Propagation is one of the most charming elements of novice radio. Radio waves' capacity to propagate via the surroundings is advocated via the usage of the sun, the ionosphere, and hundreds of atmospheric elements.

The propagation properties range in the course of frequency tiers. While VHF and UHF alerts are regularly confined to line-of-sight verbal exchange, HF indicators, as an instance, also can excursion terrific distances via bouncing off the ionosphere. Propagation can also be impacted by the time of day and the weather. For instance, higher frequency band transmissions also can journey further inside the summer season than inside the wintry climate.

How to understand Ham Radio abilities

It's important to apprehend how every HAM radio issue works if you need to operate one efficiently. The vital operations of a HAM radio's electricity supply, radio reception, radio transceiver, speaker or keyboard, and antenna are included below.

Power Source

The element that gives the radio its electric powered electricity is the energy deliver. Direct contemporary (DC) strength belongings are frequently needed for HAM radios to characteristic, and the voltage and modern-day specifications change primarily based completely on the radio's power output and exclusive additives.

Using a battery is one ordinary approach of powering a HAM radio. A commonplace opportunity is a lead-acid battery considering it's miles dependable and able to delivering a normal output of voltage and contemporary-day. But you can also use certainly one of a type types of batteries, which includes lithium-ion ones. Additionally, some HAM radios are prepared with an inner strength deliver which could convert wall-plug AC electricity into DC strength for the radio.

As variations in voltage and cutting-edge-day may want to probably harm the radio or impair its feature, it's far critical to make sure the strength supply is constant and reliable.

Radio Receiver

The radio receiver is the tool that transforms antenna-emitted signs into audio or virtual records. It is a vital a part of the radio as it affects how properly it could select out up and decipher messages.

An antenna enter, a pre-amplifier, a mixer, an intermediate frequency (IF) amplifier, a demodulator, and an audio amplifier are the standard additives of a radio receiver. The pre-amplifier amplifies the radio signal certainly so it may be dealt with through the use of the mixer as soon as it receives it at the antenna enter from the antenna. The mixer creates an IF signal with the aid of manner of mixing the amplified sign with a close-by oscillator signal, it truly is ultimately amplified through way of the usage of the IF amplifier.

The audio amplifier then amplifies the sign and feeds it to the speaker or headphones as fast because the demodulator has extracted the modulated signal from the IF sign.

Radio Transceiver

A radio transceiver combines a radio transmitter with a radio receiver. It is a important part of a HAM radio because it permits the individual to deliver and achieve indicators.

An audio input, a modulator, a strength amplifier, and an output clear out are the same old additives of a radio transmitter. The modulator is used to alternate the audio signal on the same time as it enters the audio enter from the keyboard or microphone. The audio sign and provider signal are mixed in the modulator to create a modulated signal, which is ultimately amplified via the energy amplifier.

The amplified signal is then despatched to the antenna, which transmits it after the output clear out gets rid of any undesirable alerts. One set of controls, on the side of a dial or keypad, usually operates each the radio receiver and the radio transmitter. The operator can alternate the frequency, electricity output, and distinctive parameters at the identical time as simultaneously sending and receiving statistics.

Keyboard or Speaker

The device that lets in the operator to talk with the radio is the speaker or keyboard. The operator pays interest the signs they have got obtained way to the audio output from the speaker. The operator might also additionally input data or trade parameters, which includes the frequency or energy output, using the keyboard.

To provide pinnacle audio output, the speaker or headphones must be of immoderate remarkable. The operator have in an effort to make modifications fast and precisely manner to the keyboard's responsiveness and ease of usage.

Antenna

The element that sends and receives radio waves is the antenna. It is a essential part of the radio because it controls the device's functionality for sign transmission and reception.

The antenna can be a twine dipole, a vertical, or a beam antenna, among different designs. The frequency band the operator chooses to use, further to the intended signal range and directionality, will determine the sort of antenna that is used.

To obtain effective sign transmission and reception, the antenna should be because it should be adjusted to the popular frequency range. The antenna must be tuned to healthy the frequency of the sign being sent or received via converting its length and different trends.

The positioning of the antenna is vital, in addition to tuning. Positioning the antenna as immoderate as feasible and far from any capability interference belongings, on the facet of power strains or specific antennas, is vital. To decide the nice antenna placement for their particular vicinity and frequency variety, the operator may additionally moreover want to test out numerous antenna placements.

It's crucial to recognize how every HAM radio issue works if you need to characteristic one effectively. The radio receives signs and transforms them into audio or virtual records; the radio transceiver lets in the operator to every gain and transmit signs; the speaker or keyboard permits the operator to have interaction with the radio; and the antenna transmits and receives radio signs and symptoms. The energy supply materials the electric power for the radio.

What are Ham Radio channels or frequencies?

HAM radio uses novice radio frequencies allocated with the aid of the usage of international criminal guidelines. These frequencies are separated into many frequency bands, every of which has a completely specific variety of frequencies and houses that make it suitable for sure styles of communique.

HAM radio frequency band allocations range from united states to u . S . A . But are usually given by using the International Telecommunication Union (ITU) and damaged down into many classes.

Low-Frequency Band

The LF, or low-frequency band, spans the 30 kHz to 3 hundred kHz frequency variety. Long-distance navigation and subsurface conversation are the 2 principal uses of this frequency band. Low-power transmitters and clean antennas are utilized by HAM radio operators to speak over short distances within the low-frequency band. Although this frequency variety has a confined bandwidth, interference from every natural and artificial assets can also get up.

Medium Frequency Band

Between 3 hundred kHz and three MHz is the medium-frequency band, or MF for short. This frequency band is frequently used for quick-range conversation similarly to for AM radio publicizes. HAM radio operators use extra powerful transmitters and intricate antennas to speak at some stage in greater distances inside the medium-frequency band. However, each herbal and artificial resources of interference might also affect this frequency range.

High-Frequency Band

The HF, or excessive-frequency band, spans the frequency kind of 3 MHz to 30 MHz. Since it allows lengthy-distance conversation using the ionosphere for mirrored photo and propagation, this frequency range is the most customarily applied in HAM radio transmission. Additionally, each enterprise and navy communications hire the excessive-frequency spectrum. HAM radio operators hook up with particular HAM radio operators all through the globe the usage of voice, Morse code, and digital modes like PSK31 and JT65 using the immoderate-frequency band. A form of sub-bands of the excessive-frequency band also are certain for advantageous types of verbal exchange, together with speech and information.

Very High and Ultra-High Frequency Bands

Chapter 4: How To Talk On A Ham Radio

Using a Ham radio to talk can be amusing because it permits you to set up connections with individuals anywhere inside the globe. For beginners, however, it is able to furthermore be overwhelming.

Here are the techniques you want to take to talk on a Ham Radio:

1. Get your license

A Ham radio license is wanted to use a ham radio for verbal exchange. Although the licensing machine varies from kingdom to us of a, it frequently entails passing an examination that assesses your statistics of radio principle, legal guidelines, and going for walks techniques. You can find out look at guides and exercising assessments on-line to useful beneficial useful resource together along with your exam training. After passing the check, you will be given a name sign to apply on the identical time as the usage of your radio.

2. Listen first

It's important to concentrate first to get a enjoy of what's taking location in the frequency earlier than you start sending. Get a sense for the protocols and decorum through paying attention to the dialogues. It's crucial to be privy to whilst a person is making an attempt to touch you, so maintain an ear out in your call signal.

3. Make touch

You can every name CQ (searching out a hint) or solution someone else's call on the equal time as you are prepared to set up contact. Say "CQ" after which your name symptoms and symptoms in case you are calling CQ. KI6ABC, as an example, must say, "CQ, CQ, CQ, this is calling CQ." Say the selection signal of the character whose call you are answering, then your name sign. As an instance, "KI6ABC, this is W1XYZ."

four. Introduce your self

Once you've got installed touch, pick out yourself and trade touch statistics. Your name, vicinity, and radio device want to be included on this. For example, "This is John, the usage of a Yaesu FT-857D, in Los Angeles, California."

5. Adhere to the operating guidelines

When using a Ham radio, it's far critical to stick to the working hints and protocol. Here are a few essential tips:

Wait for your turn: Don't talk over one-of-a-kind human beings at the same time as they're talking. A pause in the discourse is a incredible time to enter.

Keep it short: Keep your communications succinct and direct. Never overuse the frequency.

Be courteous: Avoid rude or contentious problems and speak in a well mannered way.

Before broadcasting, pay attention: Before broadcasting, make certain the frequency is unobstructed.

Identify your self: When broadcasting, commonly introduce your self with the useful resource of your call signal.

6. Have a communique

You can start a dialogue after changing the bare minimum of information. As long as it's far in the bounds of the going for walks rules and etiquette, you're free to speak approximately some thing you find out exciting. You are free to speak about a few element that entails thoughts, together with the climate or your hobbies.

7. End the communique

When you're prepared to stop the discussion, bid the opportunity individual farewell and sign out. Such expressions as "seventy three" (exceptional wishes) and "QRT" (getting off the air) are right. Say something like, "It changed into tremendous speaking to you, seventy three from KI6ABC."

Why is a Ham Radio crucial sooner or later of an emergency?

For survival at some point of herbal catastrophes, verbal exchange is important. When all unique way of verbal exchange have failed, Ham radio, moreover referred to as newbie radio, is critical for communique. The Federal Communications Commission (FCC) helps the usage of ham radio in instances of disaster as it's far familiar with the rate of beginner radio in such situations.

The FCC issues licenses to ham radio operators, who also can use some of frequency bands. They engage with different ham radio operators regionally, nationally, or even globally via their radios. Ham radio operators also can although touch every other using their radios in an emergency even supposing traditional verbal exchange channels like landlines, cellular phones, and the internet are not running. They may also additionally additionally provide essential records about the state of affairs, which encompass weather forecasts, evacuation warnings, and updates from emergency offerings.

Here are some motives of why using a Ham radio in an emergency is important:

1. Reliable Communication

Ham radio is a reliable form of communique at the same time as all extraordinary channels have failed. Ham radio may additionally run on batteries or mills and does now not want an external electricity supply, now not like mobile telephones or the internet. It can supply communications over terrific distances, and in superb events, even continents.

2. Independent Communication Network

Ham Radio operators be part of without relying on a centralized community. They can assemble their networks and have interaction without delay with specific operators. As a result, Ham radio is now a stand-by myself communication tool that could preserve to feature despite the reality that different varieties of communication are disrupted.

3. Flexibility

Ham radio operators can communicate with every exclusive the usage of a variety of frequencies and channels. Depending on the situations, they'll exchange frequencies or modes as crucial. For example, they could trade to a lower frequency in the direction of a typhoon to save you lightning interference.

4. Emergency services coordination

Ham radio operators can also speak with emergency offerings together with police, firefighters, and medical body of people at some stage in a crisis. They may additionally moreover offer essential statistics about the scenario and help in directing those belongings to the locations that need them the maximum.

five. Information Collection

Ham radio operators inside the impacted region would probable get essential facts approximately the state of affairs from other operators. To assist them in making options approximately a manner to deal with the hassle, this records may be shared with emergency services and distinct catastrophe response agencies.

6.   Community Support

Ham radio operators also can help their community in the course of a catastrophe. They may additionally moreover supply emotional manual to humans impacted with the useful resource of the disaster, routes to evacuation centers, and help in locating lacking parents.

When is the proper time to apply a Ham Radio

HAM radio is a beneficial device that may be applied in numerous instances, collectively with the ones requiring emergency motion. The National Incident Management System

(NIMS) is used to coordinate and control belongings in emergency reaction conditions. Emergency responders at some point of the USA use the standardized NIMS system to make certain that all agencies are speaking correctly in some unspecified time in the future of a catastrophe. With Type-1 being the maximum extreme and Type-five being the least excessive, the NIMS device divides occurrences into 5 kinds. The seriousness of the catastrophe and the best requirements of the responders decide whether or not or now not it's miles suitable to apply a HAM radio.

NIMS Type-1 Incident

The maximum vital shape of prevalence is a NIMS shape-1 incident. It consists of a huge, multi-jurisdictional response that desires an entire lot of coordination and investment. Terrorist attacks, storms, and huge-scale earthquakes are some examples of Type-1 occurrences. Communication is essential in these varieties of occurrences, and HAM radio may be a useful device.

HAM radio can be used to installation contact at some point of a Type-1 event during diverse government and jurisdictions. In places in which one among a type sorts of verbal exchange may be hampered, HAM radio operators may also moreover located up cell repeaters and base stations to offer reliable communication. HAM radio can be used to coordinate are attempting to find and rescue operations, provide situational updates in real-time, and communicate with emergency control authorities.

NIMS Type-2 or Type-three Incident

NIMS Type-2 or Type-3 incidents are greater restricted than Type-1 incidents however however call for a large response. Wildfires, floods, and amazing accidents are a few examples of Type-2 or Type-three occurrences. Communication remains crucial inside the ones kinds of emergencies, but the reaction is greater specialized, and HAM radio may not typically be required.

HAM radio may be used to installation conversation within the path of a Type-2 or Type-three event, particularly in places wherein one-of-a-kind modes of communique is probably hindered. HAM radio may be used to coordinate are seeking and rescue operations and communicate situational updates in actual-time. However, it may also be feasible to use distinct channels of communication similarly to HAM radio, like landlines and mobile telephones.

NIMS Type-four or five Incident

A small-scale reaction this is regularly managed by way of using way of 1 employer or business enterprise characterizes a NIMS Type-four or Type-5 event. Minor fires, avenue injuries, and scientific crises are some examples of Type-four or Type-five occurrences. Communication stays essential inside the ones situations, notwithstanding the fact that HAM radio won't be required.

HAM radio won't be required in a Type-four or Type-5 disaster due to the fact possibility verbal exchange techniques, which incorporates landlines and cellular phones, can be available and extra beneficial. If other modes of communication are unsuccessful, HAM radio might also additionally moreover however be applied as a fallback way of verbal exchange. Additionally, HAM radio can be used to coordinate verbal exchange among severa groups and jurisdictions, specially if the reaction calls for many businesses.

The seriousness of the disaster and the right necessities of the responders determine whether or not it is suitable to apply a HAM radio. For putting in touch among severa agencies and jurisdictions, coordinating are seeking and rescue operations, and giving actual-time updates on the state of affairs within the direction of NIMS Type-1 situations, HAM radio might be a beneficial device. Although opportunity styles of communication can be to be had, HAM radio continues to be required in NIMS Type-2 or Type-3 conditions for establishing verbal exchange and coordinating are looking for and rescue operations. HAM radio might not be required in NIMS Type-four or Type-five situations, but it may although be applied as a backup communication device and to coordinate conversation among various companies and jurisdictions.

How to combine Ham Radio into Emergency manipulate

Amateur radio is an essential device for catastrophe control. Emergency responders also can find out it hard to speak with one another on the equal time as popular verbal exchange channels are broken or overwhelmed in the direction of disasters or crises. Emergency control organizations may additionally furthermore continue to be connected and react to crises greater effectively with the useful resource of using ham radio as an alternate shape of conversation.

Chapter 5: How To Get Or Exercise For A Ham Radio License In 2023

In 2023, the gadget for obtaining or applying for a HAM radio license consists of analyzing, passing a test, and filing an software. Applying for a Ham Radio license in 2023 certifies which you are prepared to begin the usage of the provider. The beneath includes a step-through using-step guide on a manner to take a look at for a Ham Radio license in 2023.

1. Choose Your License Class

Selecting your preferred license elegance is the first step in obtaining a HAM radio license. There are 3 sorts of licenses:

Technician: This license allows using the bulk of VHF and UHF frequencies similarly to pick out HF bands with rules. The technician license is the most honest to get and is a terrific region for beginners to begin.

General: This license offers you greater rights on the majority of HF frequencies. A technician license is needed in advance than utilising for a fashionable license.

Amateur Extra: The maximum degree of privileges on all amateur radio frequencies is granted with this license. You should first private a cutting-edge license earlier than using for an beginner more license.

2. Prepare for the Test

The next step is to prepare for the test after deciding on your licensing elegance. Study property for the examination's situation rely are available online and in books. Radio jogging techniques, radio wave propagation, and radio device are a number of the topics blanketed within the take a look at.

3. Locate an Exam Session

Finding an exam session is the following step after reading for the take a look at. On its internet website, the American Radio Relay League (ARRL) continues music of approaching check dates. To discover whilst and in which check schooling are taking location, you may moreover get in contact with a nearby HAM radio membership or the Volunteer Examiner Coordinator (VEC) to your location.

four. Write the Exam

You can take the check during an examination consultation every time you are prepared. Multiple-choice questions are used within the take a look at, that is given by unpaid examiners. For the technical and desired exams, you should nicely answer at least 26 out of 35 questions, and for the amateur more examination, you need to effectively answer at the least 37 out of 50 questions.

5. Submit an Application

The next step is to submit a licensing application while you bypass the test. The FCC's Universal Licensing System (ULS) accepts programs online or thru postal mail. Your exam results want to be submitted, and a rate that varies by means of the use of license elegance have to moreover be paid.

6.   Wait in your license

The FCC will trouble your license after processing and approving your software program program. On the USAwebsite, you could look up the progress of your utility. You can begin the use of the beginner radio frequencies that your license allows you to use as speedy as it has been granted.

As in advance said, in 2023, the technique for getting a HAM radio license consists of reading, passing a check, and submitting an software program. You can earn a license that allows you to characteristic on novice radio frequencies and take advantage of the severa blessings of novice radio verbal exchange by means of using following those commands and doing all your have a have a look at. When the usage of your radio, hold in thoughts the pointers and safety precautions.

What are the styles of Ham Radio Licenses?

The Federal Communications Commission (FCC) within the US problems three incredible

HAM radio license categories: Technician, General, and Extra/Additional. The kind of license you want is based upon depend on the amount of HAM radio hobby you want to interact in. Each license has a amazing set of rights and rules.

1. Technical License

The access-stage HAM radio license is the Technician license. You are permitted to use it on all beginner bands above 30 MHz, which encompass the popular 2-meter and 70-centimeter bands. You can talk via voice, virtual modes, or maybe satellite tv for pc television for laptop verbal exchange when you have a technician license. You want to skip a 35-question multiple-choice check defensive essential radio idea, laws, and jogging techniques to get a Technician license. Morse code isn't always important for this license.

2. General License

The Technician license is followed with the aid of way of way of the General license. You can use it to function on all amateur bands below 30 MHz, which encompass the popular HF bands. You also can converse verbally, digitally, or even in morse code with a preferred license. You need to bypass a 35-question multiple-choice test defensive greater complicated radio precept, hints, and strolling strategies to get a General license. A check in five WPM (phrases steady with minute) morse code is furthermore required.

three. Extra/Additional License

The Extra license is the HAM radio license with the most privileges. It grants you the greatest degree of privileges and allows you to perform on all amateur frequencies. You can talk on all bands using voice, digital modes, and morse code if you have an Extra license. A 50-question multiple-choice check overlaying even greater complex radio principle, policies, and jogging tactics ought to be passed to get an Extra license. A take a look at in 20 WPM morse code is furthermore required.

In the usa, other licensing classes together with Novice and Advanced licenses also are to be had. The FCC no longer gives those licenses, therefore people who've already were given them also can pick out to beautify to a General or Extra license.

numerous license education exist similarly to the numerous license types, indicating the operator's degree of rights for each license kind. A technician splendor operator, as an instance, should possess a General or Extra license. Different frequency rights and running modes are approved for every license beauty.

It's critical to preserve in thoughts that getting a HAM radio license entitles you to more than truly get proper of access to to certain frequencies and conversation channels. Additionally, it includes gaining knowledge of the tips for HAM radio running and acquiring the competencies important for easy and green communication. Getting a license and attractive in the HAM radio community may be a fulfilling and informative experience, consistent with many HAM radio operators.

The Technician, General, and Extra lessons of HAM radio licenses are those which might be granted through way of the FCC in the United States. The sort of license you want depends depend on the amount of HAM radio interest you want to have interaction in. Each license has a first-rate set of rights and guidelines. In addition to reading the prison recommendations and recommendations of HAM radio going for walks and honing communication skills, getting a license gives you get admission to to excellent frequencies and communique channels.

Why is it important to have a Ham Radio license?

Some human beings might also additionally ask why it is crucial to get a Ham Radio license within the present technology with advanced communication generation like smartphones and the net. Even no matter the truth that those modern-day era clearly have their benefits, getting a Ham Radio license is essential for lots motives.

The capability to lawfully run a Ham Radio station is the number one gain of retaining a Ham Radio license. Operating a Ham Radio station without a license is unlawful and can result in outcomes or possibly prison time. In the us, Ham Radio sports are ruled thru the Federal Communications Commission (FCC), and acquiring a license demonstrates which you personal the know-how and abilities vital to run a radio station following FCC tips.

You can use diverse frequencies and communique channels that are not available to the general public when you have a Ham Radio license. HAM radios can also hire a widespread style of verbal exchange techniques, which include speech, Morse code, and digital modes, and they may feature on frequencies beginning from longwave to microwaves. When conventional avenues of communication are unavailable or unreliable, getting access to this type of big type of frequencies and communication channels is probably useful.

Ham radio operators may be a vital verbal exchange channel in emergency times. Traditional communication infrastructure, which incorporates smartphone strains and cellular towers, may additionally moreover turn out to be destroyed or overburdened inside the direction of screw ups like earthquakes, hurricanes, and wildfires, making communique hard or not possible. Ham radio operators can use their radios to hook up with certainly one of a type Ham radio operators in a few unspecified time in the future of the kingdom or perhaps the globe in similar sports, sharing essential information and coordinating treatment efforts.

Operators of ham radios can help in non-emergency times as well. For instance, they may participate in network sports like marathons or parades, helping the conversation goals of the organizers and contributing to the protection of attendees. They may also participate in radio competitions, which can be an charming and stimulating possibility to place their facts and capabilities to the take a look at.

You have to moreover make your self acquainted with the legal pointers governing radio communication generation to get a Ham Radio license. This statistics can be useful to you in various factors of your existence, like your profession or interactions with exclusive types of communique technology. You also can higher draw near how other communique generation feature thru way of analyzing radio communique.

Being a Ham Radio operator may be a worthwhile and interesting hobby further to the practical advantages of obtaining a license. Operators of ham radios frequently take part in clubs and businesses wherein they network with specific ham radio lovers and alternate facts and research. They can take part in competitions and specific sports, which can be a amusing and thrilling opportunity to place their records and talents to the check.

It's now not tough to get a Ham Radio license, and there are numerous system to help you have a observe for the take a look at. You can draw close the content cloth material and ace the test with the useful resource of have a observe substances, pattern assessments, and on-line publications. You also can observe in your license and begin taking gain of all of the perks of being a licensed ham radio operator after you pass the check.

What are the benefits of getting a Ham Radio license?

Getting a Ham Radio license has severa blessings which encompass the following:

1. Communication in an Emergency Situation

The potential to talk in times of need is one of the key advantages of keeping a Ham Radio license. Traditional conversation routes may moreover all at once emerge as overburdened or close down within the course of disasters, natural catastrophes, or super crises. Ham radio operators can reliably communicate inside the ones conditions, once in a while over fantastic distances, to provide emergency services, set up consolation operations, or deliver important data to human beings in need. You can play a essential function at the emergency reaction company in your city when you have a Ham Radio license.

2. Community Service

Many companies that offer community offerings rely upon Ham radio operators. For sports like parades, marathons, and fairs, they offer verbal exchange help, ensuring that event planners can plan sports and emergency services may additionally furthermore intervene as critical. Additionally, ham radio operators facilitate communication for community sports activities activities like festivals, carnivals, and stay indicates. You can help your network and its sports activities through way of volunteering a while and capabilities when you have a Ham Radio license.

three. Personal Communication

The opportunity to the touch one-of-a-kind Ham Radio operators all through the globe is a huge gain of shielding a Ham Radio license. Ham radio operators have interaction with every exclusive using various modes, together with speech, Morse code, digital modes, and more. Through ham radio contact, you can meet new individuals, discover one-of-a-kind cultures and international locations, and create lifetime friendships. You can contact buddies and loved ones over sizable distances with out using conventional communique channels when you have a Ham Radio license.

4. Learning Opportunities

Studying and gaining knowledge of about radio verbal exchange, electronics, and related topics are vital to get a Ham Radio license. Gaining a license may be a worthwhile instructional enjoy in and of itself, offering possibilities to broaden new talents and facts. After receiving your license, you may hold studying and experimenting with numerous modes, antennas, and equipment to hone your conversation capabilities and increase your comprehension of radio verbal exchange.

5. Experimentation

Operators of ham radios are allowed to check with numerous modes, frequencies, antennas, and quantities of bundle to hone their communication talents and boom their know-how of radio communique. You also can furthermore check with severa communication strategies, create your antennas and device, and have a study modern-day trends in the enterprise organisation with a Ham Radio license. This sorting out brought about new understandings and upgrades in the area of radio communication.

6. Career Possibilities

A Ham Radio license may additionally moreover moreover result in employment possibilities in hundreds of industries, on the facet of radio and television broadcasting, telecommunications, and emergency services. Employers understand records and capabilities consisting of conversation abilties, technical knowledge, and the functionality for self sufficient work that may be attained thru acquiring a Ham Radio license. A Ham Radio license may additionally show a dedication to education and career improvement.

7. Individual/Personal Development

An crucial private accomplishment that could offer happiness and a sense of achievement is incomes a Ham Radio license. The devotion, hard artwork, and try required to get a license can also resource in the development of individual trends like location, patience, and persistence. Furthermore, the ham radio operator network is famend for its companionship, mentoring, and help, supplying opportunities for man or woman growth and development.

eight. Enhancing Technical Knowledge

A crucial statistics of electronics, radio wave propagation, and the physics of radio transmission is essential to get a Ham Radio license. You will take a look at extra approximately how numerous antennas, transmitters, and receivers perform in addition to a way to optimize them for numerous frequencies and working modes as you got experience using your ham radio. This data may be used in numerous technological and electric powered fields, serving as a basis for studying and increase in extraordinary areas.

9. Science and Technology Exploration

In the world of radio communique, Ham radio operators have the chance to observe modern-day-day era and medical developments. As you take a look at with various tool, frequencies, and modes, you could discover new techniques for reinforcing verbal exchange over first rate distances or under tough conditions. You also can maintain up with growing technology like satellite communique, digital modes, and software program software-defined radios (SDRs).

10. Public Safety Support

Ham radio operators also can additionally help public safety businesses much like the police, fireside, and EMS similarly to supplying emergency communication in instances of catastrophe and incredible calamities. When there may be a excessive name extent or even as conventional communication channels aren't to be had, ham radio operators can also additionally moreover offer backup communication for those groups. Some newbie radio operators additionally provide their time as volunteers to CERTs and great public protection corporations.

11. International Diplomacy

Ham radio operators have a very specific danger to strengthen global cooperation and expertise. Operators may additionally additionally furthermore connect to one-of-a-kind operators worldwide, percent cultural understanding, and foster international goodwill via ham radio verbal exchange. To inspire skip-border touch and information, high quality ham radio groups moreover fund worldwide activities and exchanges.

12. Low-Cost Communication

A reasonably-priced approach of communicating over outstanding distances is thru ham radio. While some ham radio device is probably pricey, human beings on a price range have get right of entry to to a big range of much less high-priced answers. Additionally, Ham Radio communication is free of month-to-month service prices and facts use policies. You may additionally moreover speak with other operators all over the globe completely unfastened after you've got have been given your Ham Radio license and the crucial device.

thirteen.   Innovation and Invention

In the area of radio communication, ham radio operators have an extended records of invention and innovation. Ham Radio operators have been instrumental in growing the various contemporary radio communique technologies and techniques, which include frequency modulation (FM) and digital modes. Ham Radio operators may additionally deliver in this ancient past of invention and innovation thru experimenting with severa modes and gear, advancing the area of radio conversation.

Using or working a Ham Radio with out registration

Most international locations limit the use of HAM radios with out registering or acquiring a license. For instance, the Federal Communications Commission (FCC) inside the US mandates that all HAM radio operators get a license. Heavy outcomes, gadget seizure, and even jail time can be imposed for breaking this rule.

To ensure that HAM radio operators have the know-how and abilities needed to perform the tool efficiently and correctly, a license is a prerequisite. A person ought to bypass an exam overlaying subjects which includes radio idea, jogging techniques, and regulatory necessities to acquire a license.

A license can be effective for lots motives similarly to being a crook necessity. For example, a license might also permit you to be a part of a group of like-minded those who are captivated with radio communication. Additionally, it is able to offer you get entry to to a larger preference of frequencies and conversation techniques, which may be superb in a few instances.

The first step is to get a license in case you want to function a HAM radio however do no longer already have one. In the majority of countries, this includes passing an exam that gauges your familiarity with radio concept and guidelines. You may also use a number of gear, alongside side have a check manuals and exercise assessments, to aid to your exam guidance.

When using a HAM radio after receiving your license, there are some matters to endure in mind inclusive of:

First and most crucial, it is crucial to paste to all legal guidelines and guidelines governing frequency usage and strength output. This consists of obtaining any required authorizations or permits for the usage of positive frequencies or verbal exchange channels.

It's furthermore critical to be cautious of radio company interference. For instance, you should be privy to the possibility of interference with aircraft conversation structures if you are the use of a radio close to an airport. In widespread, it is useful to live some distance from operating on frequencies that might interfere with different offerings or the usage of immoderate power output.

Safety is a essential detail to recall on the equal time as the use of a HAM radio. As a part of this, ensure the equipment is securely grounded and that any electric connections are tight. Additionally, it's miles crucial to be alert to feasible dangers like excessive-voltage strength lines or lightning movements and take the critical steps to stay faraway from them.

In addition to these items preserve in thoughts, the use of a HAM radio efficaciously calls for adhering to etiquette and essential working practices. This entails identifying oneself at the equal time as speakme with other operators and the usage of the appropriate call symptoms. It moreover includes treating humans with recognize even as they are on the radio and abstaining from offensive terms and movements.

Chapter 6: Operational Techniques For A Newly Licensed Ham Radio Operator

To make sure which you use your radio successfully and correctly as a cutting-edge operator, it's far important to get acquainted with the operating protocols. Here are a few tips to get you going:

1. Know your tool: Spend a while turning into familiar along with your radio's features. Practice the usage of your radio while looking academic films and reading the commands. You will experience more comfortable the use of it and use it more successfully as a stop end result.

2. Listen first earlier than talking: Listen to the frequency to make certain no person else is the usage of it earlier than

broadcasting. Avoid interfering with every different operator's dialogue. You might also additionally examine greater approximately how your radio works in addition to the customs and conventions of HAM radio communication via listening as properly.

three. Introduce yourself: Always use your call signal to find out yourself while transmitting. In addition to being required via law, doing this is moreover polite to one-of-a-kind operators. At the start and cease of every conversation, further to as quickly as every ten mins whilst speaking, use your call signal.

4. Adhere to the guidelines: It's critical to abide through the usage of the recommendations installation by means of using the FCC and your community regulatory body. These policies guarantee each the operator's safety and the suitable usage of radio frequencies. Make positive you examine up on and apprehend the guidelines that

pertain to the frequencies you want to make use of and the license elegance you've got.

five. Be short and without delay to the component: Keep your communications brief and to the factor on the equal time as speakme. Avoid using rambling, protracted terms, and restriction the amount of non-vital talk. Keep in mind that greater operators may be searching in advance to their turn to apply the frequency.

6. Adopt suitable language and manners: There are precise languages and protocols for HAM radio communique. Using the great language and manners not first-rate improves conversation but moreover demonstrates interest for distinct operators. Always be extremely good and type, and refrain from using slang or jargon that may not be understood via each person.

7. Test your emergency communication skills: You can be requested to help in emergencies as a HAM radio operator. Practice emergency conversation protocols,

together with the use of the SOS sign, emergency frequencies, and communique with emergency response protocols.

8. Be aware about exclusive operators: Always show attention for unique clients of the frequency. Don't argue or get into heated debates, and refrain from criticizing one-of-a-type operators or their device.

nine. Keep up together collectively along with your gadget: Maintain your device regularly to assure suitable operation. This consists of doing maintenance, checking for free connections, changing batteries, and, if required, upgrading firmware or software program application.

10. Be a part of a HAM radio organisation: Think about becoming a member of a HAM radio organization. This may additionally moreover consist of signing up for a close-by HAM radio club, going to shows and competitions, and taking aspect in talk forums online. You also can additionally study greater about HAM radio, meet new people, and

feature fun the usage of your radio thru becoming concerned in a community.

What to do after calling?

Depending for your interests and the motive of the choice, there are various topics you could do after placing a HAM radio name. Here are some mind about what to do after a HAM radio call:

1. Engage in a communique

Since HAM radio is a social interest, many customers like speaking with different radio clients. A communicate about any state of affairs is viable when you region a name and call every different operator. Some operators like sharing their pastimes tour, and different interests. Others like speaking about the radio's technical workings, collectively with antennas, transmitters, or modes of operation. Always keep in mind to expose consideration for different users of the radio and test proper HAM radio protocol.

2. Exchange records

You can also trade records with the alternative operator after putting a call. Your call signal, location, signal strength, and the tool you are the use of need to all be referred to right here. In addition, you could percent more records, which consist of the weather or community statistics.

three. Take element in competitions or sports

Contests and sports activities for HAM radio are a properly-desired possibility to position your skills to the test and compete in opposition to special operators. Events and competitions are to be had in a huge shape of bureaucracy, from close by competitions to international ones. Making as many connections as you may in a certain quantity of time is needed for some sports at the same time as sharing specific records with other operators is needed for others. Meeting other HAM radio fans and honing your radio abilties can be performed thru taking element in those sports.

four. Take in special human beings's conversations

On HAM radio, you may pay attention to notable discussions even if you're not taking detail in them. This is probably an extremely good technique to get greater records on numerous subjects or to listen how different operators communicate. You can growth your listening abilties and choose up new HAM radio communique techniques via listening to specific humans talk.

5. Experiment with numerous operating modes

The more than one strolling modes supplied thru HAM radio include voice, Morse code, digital modes, and others. After putting a call, you can check out severa modes of operation to choose which of them you like. To have a have a look at how numerous frequencies or energy ranges have an impact on your sign, you can additionally test with them.

6. Join a community

A internet is a meeting of HAM radio operators that takes location often at a tough and fast frequency and time. Nets are informal meetings of operators who've comparable pursuits. They can also be installation for a selected cause, together with emergency communications. A remarkable technique to fulfill different operators and take part in institution discussions or sports is thru becoming a member of a net.

7. Perform public service

During crises or exclusive occurrences, HAM radio operators frequently provide the general public with a useful corporation. You can use your radio know-how to assist others to your network after setting a call. This might be supporting with are seeking for and rescue efforts, supporting with public event making plans, or presenting verbal exchange assist inside the route of a disaster. You also can have an first rate impact for your network and help the ones in need through using your HAM radio knowledge for public company.

Calling some other Radio station on Radiotelephony

The possibility to touch one-of-a-kind HAM radio operators the world over using various communique strategies is one of the most thrilling skills of HAM radio. Below, we can circulate over the manner to contact every special radio station the use of radiotelephony similarly to the manner to report alerts the use of each radiotelephony and radiotelegraphy (Morse code).

Calling Another Radio Station on Radiotelephony

There are some commonplace steps you have to adhere to on the identical time as the usage of radiotelephony to touch every other radio station. Always begin the call through introducing your self and the station you're phoning. For example, you could say, "CQ, CQ,

CQ, that is November Two Alpha Delta Tango calling."

You can start making your name after figuring out yourself and the station you're phoning. Generally talking, it's miles better to maintain your calls short and direct. In addition, you want to talk slowly and clearly.

The suitable way to finish a name is with the phrase "over." This we may also want to the alternative man or woman recognise that you've completed speaking and are looking for their solution. Ensure you begin speaking if the station you're phoning solutions.

Signal Reporting in Radiotelephony

An crucial part of HAM radio verbal exchange is signal reporting. Operators can check out the intensity and caliber of the sign they're getting way to it. Signal reporting is regularly finished in radiotelephony using the R-S-T scheme. For each of the following classes, quite a range of from 1 to five is given with the useful resource of this machine:

Readability: This describes the signal's clarity. A sign is absolutely illegible if it gets a rating of one, and in particular obvious if it gets a score of five.

Signal electricity is said by way of the time period "strength." A rating of one suggests a very faint sign, even as a rating of five indicates a very effective sign.

Tone: This describes the signal's tonality. A score of 1 shows signal distortion, at the same time as a rating of five suggests a smooth, distortion-loose transmission.

As an instance, you may describe a signal that have become crystal clean, very effective, and toned herbal as "5-5-five."

Signal Reporting in Radiotelegraphy (Morse Code)

Signal reporting is achieved in radiotelegraphy using a absolutely one in every of a type approach. Radiotelegraphy uses the R-S-S gadget in choice to the R-S-T method. For every of the subsequent classes, a range of from 1 to 5 is given with the useful resource of this device:

Readability: This describes how easy the Morse code is to study. Code that receives a score of one is totally incomprehensible, while code that receives a rating of five is certainly comprehensible.

Signal power is a term used to explain the signal's electricity. A score of 1 shows a completely faint signal, on the identical time as a rating of five indicates a totally powerful signal.

Tone: This describes the signal's tonality. A score of 1 denotes a distorted tone, whilst a rating of five denotes a easy, distortion-unfastened tone.

As an example, you'll probably describe a sign that become crystal clean, very powerful, and toned herbal as "599."

Directional Call in Radiotelegraphy (Morse Code)

A directed call is a radiotelegraphic technique of contacting a specific station. When many stations are the usage of the identical frequency, that is useful. You want to first pick out out the station you want to the touch to make an instantaneous name. For example, you may say, "CQ DX, that is November Two Alpha Delta Tango calling," in which "DX" denotes which you are making an prolonged-distance call.

You can start your directed call after you have were given determined the station you need to the touch. Send the station's name signal first, then your name signal, to do that. To reap a station with the selection sign "AA5AB," as an instance, you may transmit the message "AA5AB de N2ADT."

Wait for the alternative station to react once you have got got issued the directed name. If they solution, you can begin speaking to them.

General Call in Radiotelegraphy (Morse Code)

A fashionable call is used to collect every station on a fine frequency. In radiotelegraphy, you want to first discover yourself and the station you're phoning earlier than making a customary name. Using November Two Alpha Delta Tango as an example, you'll say "CQ CQ CQ, this is calling all stations."

After setting your selected call, you need to observe for any responses from stations. You might also additionally begin speakme if a station solutions.

What are the codes for a Ham Radio Operator?

Operators of Ham radio use a gadget of codes to talk with every different. These codes are intended to abruptly and efficaciously transmit data, rushing up and simplifying verbal exchange. The International Morse Code, Q codes, and RST device are only some of the codes utilized in novice radio.

Chapter 7: Organizing Your Ham Radio Home Station

It takes a few guidance and organisation enterprise to set up a domestic station for Ham Radio. Having a well-prepared station can also make running your radio extra powerful and fun, no matter whether you are an professional or beginner Ham radio operator.

1. Select a Space

Choosing the place in your tool installation is the number one diploma in putting in your private home-based totally Ham radio station. You need to preferably pick out out an area that is calm and far flung in order that no domestic sports activities will distract you from your artwork. Since you'll want room in your system and accessories, the dimensions of the region is also crucial to maintain in mind.

2. Organize your device

After selecting an area, you can begin installing region your tools. This covers any more tool you could have similarly on your radio, energy deliver, and antenna. Consider the use of cabinets, racks, or storage containers to hold topics tidy. To make it much less difficult to discover what you want, you can label your gear and accessories.

three. Set up your Work Area

The next step is to arrange your workspace. This includes the workspace in which you will use your radio, which embody a desk or workbench. Ensure that your radio, computer, microphone, and some different critical machine can all healthful in the area you have were given available. To avoid weariness throughout extended operational lessons, you may additionally need to keep in mind selecting a chair with sufficient returned manual.

4. Consider Ergonomics

The ergonomics of your workspace should be considered on the same time as designing it. This consists of making sure that your tool is set up such that it's far each stable and satisfactory so you can use. For example, you have got got to test that your chair is at the right pinnacle so you can without trouble get proper of get right of entry to to your machine. Additionally, you want to make certain that your keyboard, mouse, and microphone are set up to lessen pressure in your wrists and fingers.

5. Manage Your Cables

Organizing your Ham radio domestic station requires cautious wire management. Finding what you need is probably tough thinking about that cables are susceptible to turning into tangled and congested. Use cable ties, wire looms, or cable clips to prepare your wires. To make it less difficult to recognize your wires, you may also name them.

6. Consider Lighting

It's important to have appropriate lighting to your workspace for every comfort and protection. Make certain your workspace is well-lit, the utilization of every undertaking lighting and overhead lighting fixtures. To provide more task lighting, you could additionally need to think about the use of a table lamp with an adjustable arm.

7. Keep a Log

It's an wonderful concept to hold a magazine of your Ham radio sports activities sports to stay prepared and display your development. This can also additionally consist of maintaining a log of your contacts, device settings, and any troubles or issues you have were given. To keep song of your file, you can use a pocket ebook or pc software program software.

8. Keep it clean

It's critical to preserve a tidy and orderly domestic radio station for Hams. This entails routinely cleansing your workspace, dusting your device, and minimizing clutter. Using your radio can be extra pleasant and hold you centered if the station is neat and structured.

What are the necessities for building your Home Station

Thorough planning and thorough interest of many components are crucial at the same time as building a domestic station for Ham Radio.

Requirements for Building Your Home Station

There are some stipulations you want to do not forget in advance than beginning to layout your home station:

1. License: A modern Federal Communications Commission (FCC) license is needed to run a Ham Radio station. To get your license, you have to pass an exam that assesses your facts of radio idea, regulations, and strolling strategies.

2. Equipment: To run a Ham radio station, you want to have the right tool. This consists

of a transceiver, an antenna, a power deliver, and extra accessories.

3. Location: Your Ham radio station must be in a suitable region. This includes an area where you may installation your machine, an opening on your antenna, and get right of access to to electricity and the internet.

4. Interference: You have to make certain that your ham radio station does now not disrupt the operation of a few different digital system in your property or the close by location. Following suitable grounding and protecting tactics is part of this.

Lighting for an Amateur Radio Home Station

A stable and powerful Ham Radio station wishes right illumination. The following are a few lighting specifications to don't forget:

1. Brightness: To avoid eye pressure and to ensure you may see your device and feature a take a look at your manuals, your Ham radio station need to be nicely-lit. It is

counseled to have a brightness of at the least 50 lux.

2. Direction: To keep away from glare and shadows, your lighting fixtures need to be pointed far from your machine. To direct the slight, you may use a table lamp or a ceiling mild with color.

three. Color temperature: Your mood and productivity may be impacted via the color temperature of your lighting. For a Ham Radio station, a shade temperature of round 5000K is recommended to imitate natural sunshine.

4. Red mild: To protect their night time time time vision and lessen interference with their device, a few ham radio operators choose to use crimson lighting in their stations. Use a crimson table lamp or retrofit your current-day lighting fixtures with a red clean out.

Electrical Wiring for the Ham Radio Home Station

For your Ham radio station to run securely and dependably, right electric wiring is needed.

The following standards have to be taken underneath attention:

1. Grounding: Your Ham radio station must be well grounded to carry out properly. To keep away from electric shock and decrease interference, you need to make sure that each one of your system is effectively grounded.

2. Circuit breaker: The circuit that elements strength in your ham radio station has to have a circuit breaker or fuse geared up. In the occasion of an overload, this will defend your system and prevent electric fires.

3. Power outlet: Your Ham radio station has to have a separate energy deliver. This outlet want to be near your station and want to be rated for the amperage and voltage of your device.

four. Surge protection: To shield your device from power surges and lightning moves, you should lease surge protection gadgets. You can also install an entire-residence surge prevention tool or lease surge protectors.

What are the styles of system to apply

VHF/UHF device, HF transceivers, QRP gadget, and station accent gear are the 4 simple divisions of HAM radio machine. These companies every have special tendencies and applications in their very own.

Here are some of the system to use:

1. VHF/UHF Equipment

For brief-range communication, VHF (Very High Frequency) and UHF (Ultra High Frequency) system is carried out. While UHF uses the variety from three hundred to 3000 MHz, VHF uses the variety from 30 to 3 hundred MHz. These frequencies are used for neighborhood or nearby communique, together with that which takes area internal a town. Repeaters, satellite tv for pc television for pc communique, and mobile conversation all regularly use VHF/UHF era.

VHF/UHF Equipment Types:

Handheld radios: They are ideal for cellular communication because of the fact they may be compact and portable. They are often used for network verbal exchange, together with among hikers or on a building net site on line.

Mobile Radios: These radios are designed for use in shifting vehicles. They can broadcast over extra distances than portable radios due to the fact they may be larger and similarly powerful.

Base station radios: They are meant for utilization in a desk bound region, which incorporates a residence or an place of business. They regularly have a greater diversity and are extra effective than cell radios.

Chapter 8: Setting Up Your Pc With A Ham Radio

Your capability to talk thru radio may be appreciably improved with the useful resource of connecting your PC to a HAM radio. You can use your PC to manipulate your radio, maintain tune of contacts, or maybe decode digital transmissions if you have the essential software program software program and hardware.

The steps to installation your PC with a Ham Radio consists of the subsequent:

1. Determine Your Radio's Capabilities

You need to ascertain your radio's functionality in advance than putting in area your PC. Does your radio have a serial or USB interface just so a computer can function it? What kind of software program paintings together together together with your radio? Does your radio provide beneficial aid for digital communication strategies? You may additionally additionally determine what hardware and software program software utility you want to set up your PC via manner of responding to those questions.

2. Gather Your Hardware

After figuring out your goals, you could begin accumulating the crucial hardware. You may also want to require the subsequent hardware:

A USB or serial cable to connect your PC and radio

A sound card interface to attach your radio to the sound card on your PC.

External audio machine and a microphone for audio verbal exchange, or a pc headset.

A sign-transmitting and -receiving antenna

three. Install the software program software

Installing the software program is wanted after having all the critical hardware. For HAM radio communique, there are various software program answers available, every free and paid. Several of the popular software program program software options embody:

Ham Radio Deluxe, is a top rate software program program package deal deal that gives radio manage, logging, and digital modes.

FLdigi is free software program that could decode many digital conversation protocols.

WSJT-X is a unfastened piece of software application that makes a speciality of digital radio protocols like FT8 and FT4.

N1MM Logger is a loose logging device frequently used for competing

Follow the instructions cautiously at the identical time as putting in this gadget and putting it up to your radio.

4. Configure your radio

After this machine has been set up, you want to installation your radio for laptop control. This may additionally additionally consist of making use of a specialised computer manage software program program that the producer has given or adjusting the baud charge and one in all a type settings via the radio's menu device.

five.   Connect your hardware

You have to join your hardware after configuring your software software and radio. Utilize a USB or serial connection to connect your radio on your PC. Connect your sound card interface to each the sound card for your computer and the radio. Connect your PC to a headset, out of doors audio system, and a microphone.

6.   Test Your Setup

After connecting the whole thing, you ought to check your setup. Ensure that your radio is turned on and that the frequency and conversation mode are configured correctly. Launch your software program and make certain your radio can talk with it. To test that your audio and virtual modes are running effectively, try beginning touch with some other HAM radio operator.

7. Start talking

You can begin speakme with wonderful HAM radio operators after you have got made advantageous your machine is functioning efficiently. Use your software program software to hold tune of your contacts and QSOs. Try out numerous digital conversation strategies and employ your radio setup's superior possibilities.

What is EchoLink and its makes use of?

HAM radio operators communicate with each exclusive on line via the computer-based totally absolutely absolutely EchoLink technology. Jonathan Taylor, K1RFD, based it in 2002, and the EchoLink Software Corporation presently owns and runs it.

EchoLink hyperlinks HAM radio operators from anywhere inside the globe using VoIP (Voice over Internet Protocol). You can use your PC or a cell tool to connect with a miles off HAM radio station using EchoLink. Your computer or one of a kind device is associated with a community repeater, which is then associated with the net, for the gadget to feature. By doing so, you may engage with exclusive EchoLink-using HAM radio operators.

EchoLink is greater regularly than no longer used for emergency communications. Traditional conversation channels like smartphone strains and cellular towers can be down or overburdened in some unspecified time in the future of a herbal catastrophe or one-of-a-kind emergency situations. HAM radio operators can use EchoLink to attach and coordinate their efforts in those times.

HAM radio operators who live in locations wherein it's miles difficult to install a neighborhood repeater or who preference to talk with certainly one of a type HAM radio operators outside of their on the spot area might also moreover make use of EchoLink. Without having to put money into pricey machine, you may talk with HAM radio operators everywhere in the globe with EchoLink.

EchoLink also has severa functions that HAM radio lovers would possibly find out to be had. The generation, for instance, permits you to snoop on HAM radio nets and participate in conversations with different HAM radio operators. The device may additionally moreover help you expand your talents and take a look at from greater seasoned clients.

You want a microphone and an internet-related laptop or mobile device to make use of EchoLink. Additionally, you need to sign on an account and down load the EchoLink software. You can also additionally additionally connect with a nearby repeater and start speaking with exclusive HAM radio operators after developing an account. Using the EchoLink application, you may search for repeaters via location or call sign.

One advantage of the usage of EchoLink is how smooth and intuitive it is to set up. The software may be downloaded for gratis from the EchoLink net internet site online, and the economic enterprise offers thorough documentation and allows first-time customers. In addition, EchoLink is pretty plenty less high priced in assessment to the sometimes-pricey traditional HAM radio system.

However, the use of EchoLink has numerous drawbacks. One is that your net connection's pleasant affects the connection's first-rate. You can pay attention crackling or lose connection in case your net connection is gradual or inconsistent. EchoLink is based upon at the internet, therefore electricity outages or fantastic internet disturbances can prevent it from functioning.

EchoLink's loss of protection in evaluation to traditional HAM radio conversations is every different drawback. The system is vulnerable to hacking and splendid protection risks as it's far predicated upon on the net. As a result,

EchoLink need to now not be used to send touchy facts.

What are Hamsphere and its makes use of?

Through the net, ham radio operators may connect with distinct operators all through the globe using Hamsphere, a virtual ham radio utility. It mimics the enjoy of running a conventional ham radio station without requiring costly device or licensing.

Swedish newbie radio operator Kelly Lindman is the author of Hamsphere. Since its first release in 2010, it has gathered a following of plenty of people anywhere within the globe. For people who choice to connect to one in every of a type operators while no longer having to set up a physical station, it has grown to be a properly-favored alternative for traditional ham radio.

The software program application can be downloaded to be used with Windows, Mac, and Linux walking structures. When Hamsphere is mounted, clients may additionally additionally choose out their digital radio and frequency and then connect to particular humans at some stage in the globe to speak. Additionally, customers have the choice to format their non-public digital ham radio stations with specific antennas, power outputs, and frequencies.

One of its makes use of is obtainable to all of us with a web connection, regardless of their area or degree of license. Without the need for expensive device or a bodily radio station, it permits ham radio fans to hook up with humans everywhere within the globe.

The adaptability of Hamsphere is a in addition benefit. Voice, Morse code, virtual modes at the side of RTTY and PSK31, further to satellite television for computer tv for pc communique, are all supported. Users can also take part in competitions and other sports activities with the aid of the use of becoming a member of first rate virtual ham radio corporations.

However, utilising Hamsphere has severa regulations. It mimics the operation of a conventional ham radio station, no matter the reality that the variety and verbal exchange dependability aren't as unique. Additionally, it's far reliant on a ordinary net connection, which poses a trouble in locations with inadequate get proper of get admission to to. It does now not provide the equal diploma of emergency verbal exchange competencies as conventional ham radio, despite the fact that.

Despite the ones drawbacks, Hamsphere has received reputation as an alternative for conventional ham radio. It gives a chance for people without the sources to create a physical station to discover the interest and offers ham radio aficionados a cheap, reachable approach to talk with others across the globe. It is likewise a beneficial device for people who want to engage in competitions and awesome activities and exercising and decorate their conversation skills.

What is HPSDR?

High-Performance Software Defined Radio, or HPSDR, is an open-supply software-described radio mission that gives novice radio fans get proper of get right of entry to to to excessive-typical performance sign processing competencies. It emerge as advanced in reaction to the developing want for high-performance, reasonably priced, and really adaptable software application-defined radio systems.

Using a mixture of hardware and software software application, HPSDR offers an entire radio answer. Printed circuit boards and commonplace virtual additives are frequently used inside the layout and creation of the hardware additives through the clients themselves. On the alternative hand, the challenge's internet internet site online permits for the unfastened distribution of the software program program program additives.

An analog-to-virtual converter (ADC), a virtual-to-analog converter (DAC), and a subject-programmable gate array (FPGA) that handles real-time signal processing typically make up the HPSDR hardware. VHDL, a hardware description language this is frequently used for growing digital circuits, is used to software program the FPGA. The mixer or amplifier that gives the preferred frequency conversion and amplification is often coupled to the ADC and DAC.

On the opportunity hand, the HPSDR software application offers a hard and fast of signal-processing strategies that may be used to decode and encode numerous radio sign kinds. These algorithms, which can be designed to function with the FPGA in actual time, include filters, demodulators, and modulators. Typically written inside the C++ programming language, the software program program is meant to perform on a normal PC or computer.

The truth that HPSDR gives a excessive diploma of pliability and flexibility is one in every of its key benefits. The software can be with out trouble up to date and modified to carry out with many kinds of radio transmissions, and clients can format and convey together their private hardware components to healthful their precise desires. For amateur radio fans who need to test with numerous radio sign kinds and operational modes, HPSDR is the first-class answer.

The affordability of HPSDR is every unique gain. The radio is notably lots a whole lot less steeply-priced fashionable than conventional radio systems due to the fact the hardware additives can be constructed the usage of common digital components and posted circuit forums. Because of this, HPSDR is the proper alternative for beginner radio fanatics who need to test with diverse radio symptoms while not having to make a big financial determination.

Additionally, HPSDR could be very portable and is easy to installation and deliver in some of artwork settings. Users may moreover furthermore clearly switch the radio to extraordinary places and use it for portable or emergency communications for the purpose that software program application works on a everyday PC or computer.

The many packages for HPSDR depend on the person needs of the man or woman. HPSDR is used by some amateur radio hobbyists to test with diverse radio transmissions, together with virtual modes and willing signal modes. Since HPSDR is portable and clean to installation in lots of working settings, others placed it to use for portable or emergency communications.

HPSDR is likewise used for studies and training. It is the proper possibility for researchers and educators who need to test with numerous radio signal types and operational modes as it gives a notable diploma of versatility and flexibility. For instance, HPSDR has been implemented in educational research to take a look at how ionospheric disturbances have an effect on radio communications.

Chapter 9: Understanding Ham Radio Basics

What is Ham Radio?

Ham radio, often known as novice radio, is a shape of Wi-Fi communiqué in which individuals can transmit messages throughout various distances, from only a few miles to loads, or maybe talk with area stations. Unlike business radio or broadcasting, ham radio is precisely non-enterprise. Operators typically referred to as "hams," use numerous kinds of radio device to talk with different hams for the motive of enjoyment, know-how exchange, and public issuer.

Differences between Ham Radio and Other Communication Methods

1. Licensing and Regulation: Ham radio operators want a license to function that is acquired after passing an exam. This ensures that operators understand radio protocols, technicalities, and etiquette.

2. Frequency Spectrum: Unlike business FM/AM radios or TV declares that have everyday frequencies, hams can performed inside the course of a broader spectrum, with precise bands allocated for beginner use.

three. Nature of Communication: Ham radio is -way conversation, no longer like broadcasting it's one-to-many. Hams talk at once with every one-of-a-kind.

4. Equipment Customization: Many hams build or alter their tool. This DIY method is lots much less not unusual in precise communication techniques.

5. Purpose: While precise communication structures like mobile networks are designed for huge public communique, ham radio serves fans, emergency communications, and experimenters.

Basic Terminologies

1. Transceiver: A device which could each transmit and gather communications.

2. Frequency: The specific waveband on which communications are made.

three. Band: A range of frequencies. E.G., 20-meter band or 2-meter band.

four. Q-Codes: A standardized series of three-letter codes, like QSO (a conversation) or QTH (place).

5. DX: Refers to extended-distance conversation.

6. Repeater: A tool that gets a signal and retransmits it, extending the sort of conversation.

Advantages of Ham Radio Communication

1. Reliability: Ham radios can perform whilst other structures fail, specially in the course of natural disasters.

2. Global Reach: Depending at the frequency and conditions, hams can communicate across the world.

three. Community: Ham radio fosters a global network of fans who percent understanding, reminiscences, and assist in emergencies.

4. Learning Opportunity: It gives a practical data of radio technology, electronics, and international cultures.

five. Public Service: Many hams provide their offerings in some unspecified time within the destiny of emergencies, presenting essential communication links at the same time as favored most.

In essence, ham radio is greater than just a hobby. For many, it's far a ardour and a essential form of verbal exchange, bridging distances, cultures, and generations. Whether for public company, the satisfaction of experimentation, or the fun of creating contact, ham radio offers a few factor for each person.

Licensing and Regulations

Importance of Licensing in Ham Radio

Ham radio is a completely particular space that sits on the intersection of hobby, public company, and technological innovation. Given the ability interference with exclusive communications services and the huge spectrum it operates in, licensing ensures:

1. Operator Knowledge: A certified ham is knowledgeable approximately the technical components, working strategies, and etiquette, which guarantees smooth operation and minimal interference.

2. Safety: Licensing guarantees that operators apprehend the functionality risks of radio device, like RF exposure, and perform their stations appropriately.

three. Controlled Use of Spectrum: By regulating who can use terrific bands and frequencies, licensing prevents chaos and guarantees organized communique.

four. Public Service: In instances of emergencies, licensed operators can function a essential communication hyperlink, regularly taking detail with public carrier corporations.

Different Classes of Licenses

While the specific instructions and their privileges can variety via u . S . A ., in the United States, as an example, there are three crucial license education:

1. Technician License: This is the get proper of access to-degree license, granting privileges on all beginner bands above 30 MHz and restricted privileges in portions of the high-frequency (HF) bands.

2. General License: Building upon the Technician privileges, this license gives get right of entry to to a broader kind of frequencies and modes on the HF bands.

3. Amateur Extra License: The maximum elegance of license, granting complete privileges on all amateur bands and running modes.

Regulatory Bodies and Their Roles

Different global locations have their very very very own regulatory our our bodies overseeing novice radio operations. Their roles generally encompass:

1. Licensing: Conducting examinations, issuing licenses, and preserving a database of licensed operators.

2. Allocation of Spectrum: Designating unique frequency bands for amateur radio use.

3. Enforcement: Ensuring operators look at rules and regulations, and taking motion in competition to violations.

Chapter 10: Setting Up Your First Station

Selecting and Buying Equipment

1. Assess Your Goals: Your dreams will dictate your device desires. Are you interested by network communications, extended-distance DXing, or virtual modes? Each calls for barely one-of-a-type setups.

2. Start Simple: For beginners, a twin-band handheld transceiver is an affordable and bendy desire. As you gain experience, you may increase your setup.

3. Brands & Reviews: Brands like Icom, Yaesu, Kenwood, and Alinco are genuine within the ham radio community. Before buying, seek advice from reviews and ask fellow hams for advice.

4. Consider Used Equipment: Good awesome second-hand tools may be fee-powerful. Always take a look at used device in advance than purchase, or purchase from a trusted supply.

Basics of Antennas: Types and Their Uses

1. Dipole Antenna: The most important kind, first rate for novices. It's essentially wires pointing in opposite guidelines, with the feedline inside the center.

2. Vertical Antenna: A unmarried radio element hooked up vertically, frequently requiring a ground aircraft or radial tool.

3. Beam Antenna: Directional antennas that pay attention signs in a selected course. Yagi is a well-known type of beam antenna, specifically for DXing.

4. Loop and Magnetic Loop Antennas: They are spherical or square in form and may be compact, making them first-class for restrained areas.

five. Mobile Antennas: Designed for cellular operation, the ones are compact and frequently installation on vehicles.

Setting Up Your Station: A Step-with the aid of-Step Guide

1.   Choose a Location: Pick a quiet place with minimal digital interference. A nook of a room, a committed desk, or even a storage can art work.

2.   Set Up Your Radio: Connect the radio to a power supply. If you're the use of a base station transceiver, you may want an out of doors strength supply.

3.   Install the Antenna: If indoors, an antenna tuner can be required. For outdoor antennas, make sure they'll be immoderate up and clear of any obstructions.

four.   Connect Antenna to Radio: Use a coaxial cable to attach your antenna to the radio. Ensure connections are tight.

five.   Ground Your Equipment: This reduces interference and ensures safety. Connect your gear to a grounding rod or a suitable possibility.

6.   Test the Setup: Turn on the radio, tune to a frequency, and pay attention. You can then attempt making your first contacts!

Safety Precautions

1. Electrical Safety: Ensure all tool is properly grounded. Avoid putting in at some point of thunderstorms.

2. Climbing Safety: If mounting antennas on roofs or towers, use protection device and in no way climb by myself.

three. RF Exposure: Maintain distance from lively antennas, mainly powerful ones. Understand and comply with recommendations on regular degrees of RF publicity.

4. Equipment Safety: Regularly check device for placed on and tear. Damaged cables or device can be a fireplace danger.

five. Stay Informed: Regularly evaluation safety tips. Organizations just like the ARRL provide sources on ham radio protection.

Setting up your first station is an exciting mission into the arena of ham radio. With the right facts, tool, and safety precautions, you'll be in your manner to developing contacts and exploring the airwaves in no time!

Basic Electronics and Theory

Basic Electronic Components and Their Functions

1. Resistor: A factor used to face up to the waft of current in a circuit. It's used to govern voltage and modern-day degrees.

2. Capacitor: Stores electric strength and might launch it while wanted. Capacitors are often implemented in tuning circuits and filtering applications.

three. Inductor: A coil of twine that stores strength in a magnetic challenge while current-day flows via it. Inductors can be placed in RF circuits and transformers.

four. Transistor: A semiconductor device used to make bigger or transfer virtual signs and symptoms and electricity.

5. Diode: Allows modern-day to waft in a unmarried course however no longer the

other. They are applied in rectification and signal demodulation.

6. Integrated Circuit (IC): A set of virtual circuits on a small chip of semiconductor cloth.

Understanding Radio Waves and Frequencies

1. Nature of Radio Waves: Radio waves are a form of electromagnetic radiation with frequencies underneath 3 hundred GHz.

2. Frequency vs. Wavelength: Frequency (measured in Hertz) is the quantity of cycles of a wave that bypass a aspect in a single 2nd. Wavelength is the gap amongst one wave crest to the subsequent. They are inversely proportional.

three. Radio Spectrum: Divided into bands primarily based totally on frequency. Ham radio operates especially bands, every with its non-public homes and uses.

Modulation and Propagation Basics

1.   Modulation: The device of severa one or extra houses of a periodic waveform (issuer signal) with an facts sign. Common types in ham radio encompass AM (Amplitude Modulation) and FM (Frequency Modulation).

2.   Propagation: How radio waves journey from the transmitter to the receiver. Factors influencing propagation encompass time of day, solar hobby, and frequency.

three. Ionospheric Layers: The ionosphere has layers that can mirror radio waves, contemplating lengthy-distance verbal exchange. Layers like the D, E, and F layers play a key characteristic in HF propagation.

Basic Circuit Diagrams and Their Significance

1.   Purpose of Circuit Diagrams: They offer a visible instance of the way digital components are connected, making it less complicated to recognize, format, or troubleshoot circuits.

2. Symbols: Each component (like resistors, capacitors, or transistors) has its very very own standardized photo.

three. Reading Diagrams: Learn to trace paths of circuits, choose out components, and understand the go with the glide of modern-day-day.

four. Importance for Hams: Hams frequently alter or restore their system. Being able to look at and apprehend circuit diagrams is essential for the ones obligations.

Understanding the fundamentals of electronics and radio precept is foundational for genuinely all people diving deep into ham radio. It's no longer pretty much running the machine, but know-how the technological information and ideas in the returned of it, which leads to higher operation, troubleshooting, and a deeper appreciation for the interest.

Operating Techniques and Procedures

Making Your First Contact

1.   Listen First: Spend a while tuning across the bands and listening. Get a experience for the continuing conversations and protocols.

2.   Choosing the Right Frequency: Ensure the frequency you are interested by is not being used. It's polite to invite, "Is this frequency in use?" in advance than starting a conversation.

three. Initiate a Call: Start with "CQ CQ CQ" accompanied through manner of your name signal. For example, "CQ CQ CQ from [Your Call Sign]."

four. Making a QSO (Conversation): Once someone responds to your call, speak in reality, and trade relevant info like signal report, place, and device information.

Q-codes and Their Uses

1.   Purpose of Q-codes: Q-codes are 3-letter codes that originated from Morse code communications. They condense common

questions or statements into quick, standardized codes.

2. Common Q-codes:

QSO: A conversation with every other ham.

QTH: Location.

QRZ: Who is looking me?

QSY: Change frequency.

QSL: Confirmation receipt.

3. Using Q-codes: While they originated in Morse (CW) operations, Q-codes also are usually applied in voice operations for brevity.

Phonetic Alphabet

1. Purpose: To make sure readability all through communication, mainly at the same time as sign conditions aren't principal.

2. Standard Phonetic Alphabet:

Chapter 11: Advanced Operating Techniques

Satellite Communication Using Ham Radio

Overview: Amateur radio satellites, frequently referred to as "ham satellites," offer hams a completely unique possibility to speak thru place-based totally absolutely repeaters.

1. Basics:

Uplink: The frequency on that you transmit to the satellite tv for laptop.

Downlink: The frequency on which you get preserve of signs from the satellite tv for computer.

2. Types of Satellites:

Low Earth Orbit (LEO) Satellites: These are close to Earth and journey in an orbit fast. They are available with essential machine however are exceptional internal range for a short length.

Geostationary Satellites: These stay in a difficult and speedy function relative to some extent on Earth and provide ordinary coverage to a particular place.

three. Equipment and Antennas: While some satellites can be accessed using a dual-band handheld radio and a hand held directional antenna, extra advanced setups might also require rotators to tune the satellite, and extra powerful transceivers.

four. Making a Contact: The Doppler impact will exchange the frequency of the satellite tv for pc sign as it actions relative to you. Adjusting for that is vital.

Digital Modes and Software-Defined Radio (SDR)

Overview: Digital modes encode voice or records into virtual symptoms, transmitted over the airwaves. SDR uses software application to carry out the numerous talents historically managed via hardware.

1. Popular Digital Modes:

RTTY (Radio Teletype): An early form of virtual verbal exchange.

PSK31: Known for its efficiency and slow-velocity communication.

FT8 & FT4: Modes designed for vulnerable sign communication.

2. Software-Defined Radio:

SDR permits for flexibility, as changing the radio's behavior is as easy as updating software program.

Popular SDR structures consist of the FlexRadio series and RTL-SDR.

3. Integration with Computers: PCs can decode digital signals using software along side WSJT-X, FLdigi, and Ham Radio Deluxe.

Moonbounce (EME) and Meteor Scatter Communication

Overview: These strategies contain bouncing radio alerts off natural celestial our our bodies or meteor trails.

1. Moonbounce (EME: Earth-Moon-Earth):

This includes sending signs to the moon and receiving the contemplated sign yet again on Earth.

Requires excessive strength, large antennas, and precise positioning.

2. Meteor Scatter:

Uses the ionized trails of meteors getting into the Earth's environment to mirror radio signs.

Short conversation bursts may be completed at the equal time as situations are proper.

Portable and Mobile Operations

Overview: Ham radio isn't confined to a home station. Many hams take their system on the street or to far off locations.

1. Mobile Operations:

Operating from a automobile. This requires cellular transceivers, antennas designed for car mounting, and strength sourced from the auto.

Challenges encompass noise interference from the automobile and confined antenna duration.

2. Portable Operations:

Operating from transient locations like parks, mountains, or in some unspecified time in the future of place days.

Equipment is often battery-powered. Portable antennas and masts is probably used.

Popular sports activities encompass Summits on the Air (SOTA) and Parks on the Air (POTA).

Advanced jogging strategies in ham radio offer a captivating combination of radio era, software program application engineering, and bodily demanding conditions. While a number of those sports activities sports could probable require funding in tool and time, they open up avenues for communique and experimentation which can be tremendous inside the international of novice radio.

Troubleshooting and Maintenance

Common Problems and Solutions

1. No Power:

Check Power Supply: Ensure the energy supply is plugged in and became on.

Fuse: Examine the fuse. Replace if blown.

2. Low or No Audio:

Volume Control: Confirm the amount isn't thru coincidence grew to turn out to be down.

External Speaker: If the usage of one, make certain it's efficaciously related.

Headphone Jack: Occasionally, dirt or harm can purpose issues.

three. Poor Reception:

Antenna Connection: Confirm the antenna is securely related.

Band and Frequency: Ensure you're on the appropriate band and frequency.

Interference: Household electronics can now and again motive interference. Identify and restrict belongings.

4. Transmitting Issues:

Microphone: Ensure it's efficiently associated and functioning.

SWR (Standing Wave Ratio): A high SWR shows an problem with the antenna system. Adjust or take a look at for damage.

Routine Maintenance and Checks

1. Visual Inspection:

Periodically take a look at all machine for visible harm or placed on.

2. Cleanliness:

Dust off machine often.

Clean connection elements to make certain most incredible touch.

three. Antenna Inspection:

Check for bodily harm, in particular after immoderate weather.

Verify that mounting factors are consistent.

4. Connection Tightness:

Over time, connections can loosen. Periodically tighten antenna, ground, and electricity connections.

five. Battery Checks:

If using battery-powered system, check the battery's fitness and charge frequently.

Advanced Troubleshooting Techniques

1. Use of Multimeter:

Measure voltage, present day, and resistance to come to be aware of troubles.

2. SWR Meter:

Use to affirm the antenna tool is going for walks effectively.

3. Spectrum Analyzer:

Helps visualize and find out troubles with sign frequency, interference, and power.

four. Signal Tracing:

Use an oscilloscope to trace and think about sign paths, assisting to pinpoint wherein a signal might be misplaced or degraded.

Safety When Troubleshooting

1. Turn Off and Unplug: Before strolling on any device, turn it off and unplug it.

2. Avoid Liquid: Ensure the workspace is dry. Liquid can pose a danger of brief circuits and electric powered surprise.

three. High Voltage Precautions: Some gadget, specifically tube-based definitely tools, can preserve immoderate voltages despite the fact that grew to end up off. Be careful.

4. Use Proper Tools: Using the right equipment for the device can save you unintended damage and make sure safety.

5. Grounding: Ensure your station and tool are properly grounded to decrease the chance of electrical wonder and to enhance average performance.

6. Stay Informed: Read the guide. Understand the gadget's safety hints and adhere to them.

Regular preservation ensures the durability of ham radio tool and complements general overall performance. When problems do rise up, a methodical approach to troubleshooting can efficiently choose out and remedy issues. Always prioritize safety, ensuring that troubleshooting and preservation sports do no longer endanger the operator or others.

Safety Practices in Ham Radio Operations

Safety is paramount in any technical area, and ham radio isn't any exception. The following protection practices make certain that operators continue to be secure whilst installing place, on foot, and preserving their ham radio device.

1. Electrical Safety:

Disconnect Power: Always disconnect energy earlier than working on any system.

Use Proper Fuses: Make high-quality to use the encouraged fuse for every device to save you overload.

Avoid Water: Keep all electric gadget dry and do not feature radios with wet hands or in damp environments.

Ground Equipment: Properly ground all ham radio device to prevent electric powered powered shocks and to lower interference.

2. Antenna Safety:

Location: Ensure antennas are positioned at safe distances from strength traces, trees, and specific obstructions.

Maintenance: Regularly take a look at antennas for placed on and tear. Make maintenance or replace additives as essential.

Weather Precautions: Lower antennas inside the route of storms, specifically if there may be a risk of lightning.

Installation: When erecting antennas, usually have a spotter or assistant to make sure protection.

three. Equipment Safety:

Ventilation: Ensure that radio gadget has top sufficient air go with the glide to save you overheating.

Cable Management: Organize cables to keep away from tripping dangers and preserve them a long way from warmth property.

Regular Inspection: Check gadget periodically for placed on, tear, or defects.

4. RF (Radio Frequency) Exposure Safety:

Limit Exposure: Minimize the time spent in the the front of the antenna while transmitting.

Awareness: Be aware of the RF publicity limits set thru regulatory bodies and make sure you do no longer exceed them.

Protect Bystanders: Ensure family individuals, pals, or bystanders aren't exposed to immoderate ranges of RF power.

five. Portable and Mobile Operation Safety:

Secure Equipment: Ensure mobile tool in motors is securely mounted to avoid motion all through unexpected stops or injuries.

Avoid Distractions: If on foot at the same time as using, prioritize using protection. It's regularly satisfactory to pull over if prolonged communique is needed.

Battery Safety: When working portably, use the suitable battery kinds and make certain they will be successfully ventilated.

6. General Safety:

Stay Informed: Regularly examine safety recommendations and attend workshops or seminars on ham radio safety.

First Aid: Keep a primary useful aid package deal deal reachable in your ham shack or on foot location.

Protect Your Hearing: Be careful approximately quantity stages, in particular even as the usage of headphones.

Safety in ham radio operations is a non-prevent machine. Being vigilant, informed, and usually erring on the side of warning will ensure a safe and interesting enjoy for all radio enthusiasts.

Joining the Ham Radio Community

Local Clubs and Organizations

1. Benefits of Joining:

Access to shared resources and gadget.

Learning from skilled contributors.

Participation in organized activities and sports sports.

2. Finding Local Clubs:

The ARRL (American Radio Relay League) or its equal in unique international places often keeps a directory.

Local electronics or hobby stores may also additionally furthermore have records.

Attending ham radio conventions or change meets can introduce you to neighborhood golf equipment.

3. Participation:

Attend regular conferences.

Engage in membership sports activities and volunteer for management roles or event company.

Ham Radio Events and Field Days

1. Field Days:

Annual activities in which hams workout emergency response skills.

Set up transient stations, frequently powered thru generators or solar panels.

Focus on making as many contacts as possible in a precise duration.

2. Hamfests and Conventions:

Gatherings of hams to buy, promote, or trade device.

Often function presentations, workshops, and licensing training.

3. Special Event Stations:

Stations set up to commemorate unique activities or anniversaries.

Offer unique opportunities to accumulate special QSL cards or certificate.

Online Forums and Resources

1. Popular Forums:

QRZ: A massive database of name symptoms and signs and symptoms and signs and a famous discussion board.

eHam: Features opinions, forums, and articles.

Ham Radio Deluxe: Forums related to the famous software program software suite.

2.　Learning Resources:

YouTube: Many hams have channels devoted to tutorials, opinions, and ham sports activities.

Blogs: Numerous beginner radio fans preserve blogs detailing their adventures and duties.

three. Social Media:

Platforms like Facebook, Twitter, and Reddit have lively ham radio groups. Joining agencies or following applicable hashtags can be useful.

Building a Network and Mentorship

1.　Elmering:

In ham jargon, a mentor is referred to as an "Elmer." Seek out professional hams willing to manual green parents.

Learn from their reviews, ask questions, and trying to find recommendation on advancing within the hobby.

2. Collaborative Projects:

Engage in institution builds or antenna raising days.

Join or provoke group expeditions for SOTA or IOTA (Islands at the Air) activations.

three. Regular Nets:

Nets are scheduled on-air conferences on unique frequencies. Participate regularly to construct connections and stay knowledgeable.

Examples encompass emergency preparedness nets, unique interest nets (like sailing or RV traveling), and local check-in nets.

Joining the ham radio network enriches the enjoy of the interest. The shared facts, camaraderie, and collective ardour for radio communication offer now not handiest a technical basis but also foster lifelong friendships and a sense of belonging in a global community. Whether offline or online, every interplay is a step in the direction of becoming a more protected and lively member of this colourful community.

Chapter 12: Communicating With Exclusive Hams

Communicating with different hams, often termed "making contacts," is one of the maximum profitable factors of being an novice radio operator. Whether it's miles a short network chat or an extended conversation with someone from a distant u . S . A ., every interplay contributes for your skillset and enriches the ham enjoy. Here's a guide to creating and fostering contacts:

1. Starting Locally:

Repeater Networks: Many hams begin through manner of getting access to network VHF and UHF repeaters. These are gadgets that get maintain of transmissions from a handheld or cellular radio after which retransmit them on a very precise frequency, extending their variety.

Net Check-ins: Many repeater agencies have scheduled "nets" where operators can test in, percentage data, and workout emergency communique drills.

2. Expanding Your Range with HF:

After acquiring General or Extra beauty privileges, you could get proper of access to the HF bands, which permit for neighborhood, country wide, and global conversation.

Calling CQ: This is the equal vintage manner to initiate a conversation. By transmitting "CQ CQ CQ" found via your call sign, you're basically inviting any listening station to reply.

three. Special Modes and Methods:

Digital Modes: Technologies like FT8, PSK31, and RTTY allow for communique using virtual alerts, often permitting contacts even under negative band conditions.

Satellite Contacts: Hams can talk via satellites specially designed for novice radio. This calls for records the satellite tv for pc tv for computer's skip instances and frequencies.

Moonbounce (EME): Advanced operators may even soar signals off the moon to speak with others.

4. DXing:

The term "DX" refers to far flung stations. DXing is the interest of organising contacts with some distance-off stations, mainly ones specifically international locations.

Many hams experience chasing "DX entities" and trying to verify contacts with as many countries or regions as viable.

five. Engage in Contests:

Radio contests are events wherein hams attempt to make as many contacts as feasible indoors a fixed time body. They're a extremely good manner to decorate your running talents and make numerous contacts.

Some contests consciousness on particular modes, bands, or areas.

6. QSL Cards and Confirmations:

After growing a hint, hams frequently trade QSL gambling cards, which is probably postcards that verify the contact's details.

Electronic confirmation structures like Logbook of The World (LoTW) and eQSL have grow to be well-known for that reason.

7. Be Respectful and Patient:

Listen earlier than transmitting to make sure you're not interrupting an ongoing verbal exchange.

Always use your name sign to end up aware of your self, and be courteous and affected individual, in particular with new operators or those whose primary language can also fluctuate from yours.

8. Join Clubs and Groups:

Being a part of a nearby or hobby-primarily based absolutely ham radio club can provide countless opportunities to research, perform unique occasion stations, and interact in employer sports.

Clubs frequently have "Elmers" or mentors who can manual rookies.

nine. Participate in Events:

Events like Field Day, JOTA (Jamboree at the Air), or particular event stations commemorate unique sports and provide specific possibilities for making contacts.

10. Continuous Learning:

The international of ham radio is sizable and always evolving. As you talk with particular hams, live curious, ask questions, and look at from every interplay.

Remember, each ham changed into as quickly as a newbie, so do now not be intimidated. Dive in, begin making contacts, and enjoy the widespread and welcoming worldwide of novice radio!

Getting Your Ham Radio License

Getting your ham radio license, regularly referred to as an newbie radio license, is a profitable project that opens up a global of verbal exchange opportunities. Here's a step-with the aid of-step guide on the way to advantage your license:

1. Decide Which License Class to Pursue:

There are typically 3 schooling of newbie radio licenses:

Technician Class: This is the access-degree license. It offers you all VHF/UHF beginner bands and some HF privileges.

General Class: This is the intermediate stage. It allows get right of entry to to more frequencies on the HF band, increasing your capacity to speak globally.

Amateur Extra Class: This is the excellent stage of license, granting privileges on all beginner bands and modes.

Most newbies start with the Technician magnificence and improvement to better education over time.

2. Study for the Exam:

Obtain take a look at substances mainly tailor-made for your selected license elegance.

Consider numerous assets:

Books just like the ARRL Ham Radio License Manual

Online check courses and guides

Mobile apps for examination education

Local membership classes

three. Take Practice Exams:

Several internet sites provide loose exercise checks. These can help gauge your readiness and grow to be privy to areas needing extra evaluation.

Questions in those exams are from the first rate query swimming swimming swimming pools, in order that they closely mirror what you'll stumble upon inside the actual check.

4. Find an Exam Session:

The ARRL internet website on-line (for the ones within the US) has a searchable database of upcoming examination lessons.

Exams are frequently administered with the aid of neighborhood novice radio clubs or volunteer examiners.

five. Take the Exam:

Bring a picture ID and any required charges to the exam session. The rate typically covers the rate of administering the exam.

If you are looking to enhance to a better license elegance, deliver a reproduction of your cutting-edge-day license.

Exams are multiple-choice. You'll need to get a minimum variety of questions proper to skip (e.G., 26 out of 35 for the Technician beauty).

6. Wait for License Grant:

After passing, your examination outcomes are sent to the FCC (in the US).

You'll be assigned a unique name sign. Once it seems inside the FCC's Universal Licensing System (ULS) database, you're formally licensed and might begin transmitting.

7. Join the Ham Community:

Consider becoming a member of close by novice radio clubs. They regularly offer mentorship for inexperienced parents.

Participate in activities like Field Day, which offer arms-on enjoy and possibilities to test from pro operators.

eight. Continuous Learning:

The global of ham radio is tremendous. Stay curious and maintain to find out about one-of-a-kind modes, bands, and era.

Consider upgrading your license as you come to be extra gifted. Each enhance expands your privileges and potential for brand spanking new stories.

Remember, ham radio is as a great deal about community as it is approximately conversation. Engage with fellow lovers, are looking for steerage while favored, and constantly perform within the guidelines and spirit of newbie radio.

Federal Communications Commission

The Federal Communications Commission (FCC) is the authorities commercial company corporation responsible for regulating interstate and international communications with the aid of radio, television, wire, satellite tv for pc tv for laptop, and cable within the U.S. One of its many responsibilities includes the licensing and law of beginner (ham) radio operators.

Ham Radio Licensing by using the usage of the use of the FCC: An Overview

1. Purpose of Licensing:

The primary aim is to make sure that ham radio operators apprehend radio protocol, can carry out gadget as it should be, and are informed approximately the guidelines governing the airwaves.

Licensing moreover prevents interference with other vital communication services.

2. License Classes:

The FCC currently troubles three specific training of newbie radio licenses:

Technician: This is the get entry to-degree license. It provides get entry to to the VHF/UHF bands and some confined HF (shortwave) privileges.

General: This splendor offers broader get right of entry to across the HF bands, contemplating worldwide communications.

Amateur Extra: This most level of license offers all to be had privileges on all bands and modes.

Each better elegance license includes passing a regularly greater hard exam, shielding every radio concept and the FCC's pointers for beginner radio.

3. Examination:

Exams are administered with the aid of using manner of Volunteer Examiners (VEs), famous by way of using Volunteer Examiner Coordinators (VECs).

The assessments embody multiple-choice questions taken from publicly to be had question swimming pools.

The Technician and General assessments every have 35 questions, on the equal time because the Amateur Extra examination has 50.

These swimming swimming swimming pools are periodically reviewed and up to date to reflect cutting-edge technology and guidelines.

4. License Validity and Renewal:

Once issued, an novice radio license is legitimate for ten years.

Licenses can be renewed on-line thru the FCC's Universal Licensing System (ULS) or via mail. There's no check required for renewal.

If now not renewed in the grace length (years after expiration), the license and talk to sign are canceled.

5. Call Signs:

Every licensed novice radio operator gets a very precise name sign, used to discover oneself at the air.

The FCC robotically assigns sequential name symptoms and signs and symptoms, however operators also can request specific "arrogance" name signs and symptoms, provided they'll be available.

6. Operating Regulations:

Ham radio operators want to paste to particular suggestions referred to with the aid of way of the FCC, at the side of wherein (which frequencies/bands) they're capable of characteristic, the maximum energy tiers, and the modes they are able to use.

The specifics range primarily based on the license beauty.

7. Violations and Enforcement:

The FCC video display devices the airwaves and investigates evaluations of interference, unlicensed operation, and awesome violations.

Violations can bring about fines, license revocation, or top notch outcomes.

eight. Additional Responsibilities:

Beyond licensing, the FCC has specific responsibilities like allocating spectrum, dealing with interference proceedings, and making sure the overall orderly use of the airwaves for all communication services, not just newbie radio.

To keep abreast of the today's necessities, guidelines, and techniques, it's miles an exquisite concept for ham radio operators to periodically assessment the applicable sections of the FCC's guidelines or talk over with reliable guides from agencies much like the American Radio Relay League (ARRL).

Chapter 14: Studying For The Ham Radio Technician Exam

Studying for the Ham Radio Technician examination requires a combination of expertise theoretical information and fingers-on enjoy. Here's a manual to help you efficaciously prepare:

1. Obtain the Right Study Materials:

Official Manuals: Purchase or borrow an reputable have a look at guide or guide tailored for the Technician beauty license, which incorporates the "ARRL Ham Radio License Manual."

Question Pools: Access the modern-day query pool for the Technician exam. The real examination questions are drawn from the ones swimming swimming pools.

2. Schedule Regular Study Sessions:

Dedicate specific times for studying to ensure you constantly cover the cloth.

Break your study time into capability chunks, interspersed with short breaks, to keep consciousness and retention.

three. Cover the Basics:

Familiarize your self with the essential components of amateur radio, which includes frequencies, primary operations, and etiquette.

Understand number one electronics and radio wave houses.

4. Active Learning Techniques:

Summarize Each Section: Once you finish a financial ruin or section, summarize the principle factors in your non-public phrases.

Flashcards: Use flashcards to test your understanding. One component ought to have a query or time period, and the opposite difficulty the solution or definition.

Teach Someone: Explaining what you've got determined out to a person else, regardless of the reality that it's miles just an imaginary scholar, can assist consolidate your expertise.

5. Practice Exams:

Numerous on-line structures provide loose workout assessments for the Technician elegance license.

Regularly take those assessments to gauge your preparedness and find out weak areas.

Review the reasons for any questions you pass over.

6. Join a Local Ham Radio Club or Class:

Many clubs provide courses or take a look at companies for people getting equipped for the Technician examination.

Interacting with experienced ham radio operators can provide valuable insights and clarifications.

7. Get Hands-on Experience:

If possible, go to a ham radio operator's station or attend a "Hamfest."

Familiarize yourself with radio system, antennas, and easy operations.

8. Stay Updated:

Regulations, generation, and practices in ham radio might probably evolve. Ensure your have a observe substances are up to date.

Join online forums or groups, like QRZ.Com, to stay knowledgeable and get help.

nine. Review and Self-verify:

Regularly skip again and assessment sections you determined difficult.

Use exclusive property or strategies for topics you discover especially difficult to recognize.

10. Stay Relaxed and Positive:

The day earlier than your examination, loosen up and get a notable night time's sleep.

Approach the test with self belief, understanding you've got organized thoroughly.

Remember, the purpose isn't always absolutely to skip the exam but to apprehend and experience the location of ham radio. With self-control and the proper technique, you may now not great pass the take a look at however furthermore embark on an thrilling journey inside the amateur radio community.

Sample Test Questions and Answers

Questions Based on Each Chapter

1. Understanding Ham Radio Basics:

Which of the following amazing describes the number one motive of ham radio?

A) Commercial broadcasting

B) Professional emergency communications

C) Licensed shape of non-enterprise verbal exchange for hobbyists

D) Unlimited worldwide broadcasting

2. Licensing and Regulations:

In the U.S., what organisation oversees ham radio licensing?

A) ARRL

B) NASA

C) WHO

D) FCC

three. Setting Up Your First Station:

Which trouble is maximum critical for each receiving and transmitting alerts in a ham radio setup?

A) The speaker

B) The strength deliver

C) The antenna

D) The display show

4. Basic Electronics and Theory:

Which tool is used to save electric powered powered fee fast?

A) Resistor

B) Transistor

C) Inductor

D) Capacitor

5. Operating Techniques and Procedures:

What does the Q-code "QSL" normally advocate in ham radio communication?

A) Change your frequency

B) Acknowledgment or confirmation of receipt

C) End of communication

D) Emergency situation

6. Advanced Operating Techniques:

Which communique approach entails bouncing alerts off the moon's floor?

A) Tropospheric scatter

B) Meteor scatter

C) NVIS

D) Moonbounce (EME)

7. Troubleshooting and Maintenance:

If there can be an unusually excessive SWR reading on your transceiver, what should be the number one aspect to look into?

A) The keypad

B) The electricity deliver

C) The microphone

D) The antenna device

8. Joining the Ham Radio Community:

An professional ham radio operator who publications and enables novices is often referred to as?

A) Guide

B) Instructor

C) Elmer

D) Leader

Detailed Answers with Explanations

1. C - Ham radio is usually a certified form of non-commercial business enterprise conversation for hobbyists and fans.

2. D - The Federal Communications Commission (FCC) oversees ham radio licensing within the U.S.

3. C - The antenna is vital for each receiving and transmitting signals in a ham radio setup.

4. D - A capacitor is used to save electrical rate in quick in a circuit.

5. B - "QSL" is a Q-code that stands for acknowledgment or confirmation of receipt.

6. D - Moonbounce, or EME (Earth-Moon-Earth), includes bouncing signs off the moon's floor.

7. D - If there's a immoderate SWR analyzing, the number one element to test

out have to be the antenna tool, because it suggests capability mismatches or issues there.

eight. C - An skilled ham radio operator helping beginners is often known as an "Elmer."

Tips for Taking the Test

1. Preparation: Start getting equipped properly in advance of your take a look at date. Consistent have a look at through the years is greater powerful than cramming.

2. Mock Tests: Take mock exams to get familiar with the query pattern and to assess your preparedness.

three. Stay Calm: Nervousness will have an impact on common performance. Take deep breaths and live snug.

four. Read Carefully: Ensure you study each query cautiously in advance than answering.

five. Time Management: Keep an eye on the clock, ensuring you do now not spend too much time on any single query.

6. Stay Updated: Regulations and requirements may in all likelihood trade. Ensure you are analyzing the most modern-day fabric.

7. Join Study Groups: Studying with others can provide precise views and make clear doubts.

8. Review: After finishing the check, if time lets in, evaluation your solutions to make certain you haven't made any apparent errors.

Arming yourself with information, training regularly, and following those guidelines will boom yourself guarantee and possibilities of passing the ham radio test.

20 extra pattern take a look at questions and solutions:

1. What is the primary frequency band for lengthy-distance shortwave radio communication?

A) VHF

B) UHF

C) HF

D) LF

Answer: C) HF

2. In which type of modulation is the amplitude of the company wave numerous in accordance with the signal's amplitude?

A) FM

B) AM

C) SSB

D) CW

Answer: B) AM

3. Which aspect converts electric powered electricity into sound waves in a radio set?

A) Transistor

B) Diode

C) Speaker

D) Resistor

Answer: C) Speaker

four. What is the meaning of the Q-code 'QRZ'?

A) Who is looking me?

B) Can you renowned receipt?

C) Shall I increase energy?

D) Are my signs fading?

Answer: A) Who is asking me?

five. Which of the following describes a dipole antenna?

A) A single vertical pole

B) Two factors producing and receiving alerts

C) A spherical antenna

D) An antenna with a couple of factors above a floor aircraft

Answer: B) Two factors generating and receiving indicators

6. What type of wave consists of radio indicators amongst transmitting and receiving stations?

A) Sound waves

B) Gamma rays

C) Electromagnetic waves

D) Seismic waves

Answer: C) Electromagnetic waves

7. In what state of affairs could you use the 'Morse code' distress sign 'SOS'?

A) Starting a verbal exchange

B) Ending a communication

C) Emergency situation

D) Testing tool

Answer: C) Emergency state of affairs

eight. What is a repeater used for in ham radio?

A) To record indicators

B) To boom the fashion of transmissions

C) To exchange frequency

D) To decode digital signs

Answer: B) To growth the range of transmissions

9. Which ham radio license class is the maximum advanced within the U.S.?

A) Technician

B) General

C) Amateur Extra

D) Novice

Answer: C) Amateur Extra

10. What determines the frequency of a radio wave?

A) Wave top

B) Wave width

C) Wave speed

D) Wave period

Answer: D) Wave length

11. Why is grounding vital to your ham radio station?

A) To enhance reception

B) For safety and to save you interference

C) To boom transmission strength

D) To cool the device

Answer: B) For safety and to prevent interference

12. Which a part of a radio wave remains ordinary and is used to preserve the records?

A) Amplitude

B) Frequency

C) Phase

D) Carrier

Answer: D) Carrier

thirteen. What is the reason of a 'balun' in an antenna device?

A) To balance unbalanced loads

B) To decode signs

C) To increase susceptible alerts

D) To filter out noise

Answer: A) To stability unbalanced hundreds

14. What is 'DXing' in the context of ham radio?

A) Fixing tool

B) Transmitting digital signals

C) Contacting distant stations

D) Broadcasting information

Answer: C) Contacting remote stations

15. Which form of propagation lets in radio waves to journey past the horizon, in particular on HF bands?

A) Ground wave

B) Skywave

C) Line of sight

D) Direct wave

Answer: B) Skywave

16. What does the term 'Rig' speak with in ham radio lingo?

A) The license

B) The operator

C) The station system

D) The frequency

Answer: C) The station gadget

17. Which tool may be used to diploma SWR in an antenna device?

A) Oscillator

B) Amplifier

C) Transceiver

D) SWR meter

Answer: D) SWR meter

18. Which of the following codes means 'Your sign is fading'?

A) QSB

B) QSL

C) QRZ

D) QRV

Answer: A) QSB

19. On which of the following frequencies is the 2-meter band by means of the use of and large positioned?

A) a hundred and forty four-148 MHz

B) 430-440 MHz

C) 50-fifty four MHz

D) 7.Zero-7.Three MHz

Answer: A) a hundred and forty 4-148 MHz

20. Which of the following is NOT a shape of modulation?

A) AM

B) FM

C) PM

D) BM

Answer: D) BM

Beyond the Test – Continuous Learning and Exploration

After passing the take a look at and obtaining your license, your ham radio adventure may want to not surrender; it truely starts offevolved. This monetary destroy outlines the course of continuous boom, wherein the actual adventure and amusing of ham radio lies.

1. Upgrading Your License

a. The Importance of Upgrading: Understand the advantages, elevated privileges, and opportunities that include better-class licenses.

b. Steps to Upgrade: From locating take a look at property to taking the following-diploma take a look at.

c. A Glimpse into Advanced Licensing: A have a examine the greater spectrum privileges and obligations of better license commands.

2. Experimenting with Advanced Projects

a. Building Your Own Equipment: Dive into DIY with kits, schematics, and developing custom radio tools.

b. Antenna Design and Construction: Explore various kinds of antennas, their designs, and advantages.

c. Digital Modes and Software-Defined Radios: Delve into the future of ham radio with digital communique techniques.

d. Remote Station Operation: Understand the technology and advantages within the returned of remotely operating your station.

3. Teaching and Mentoring New Ham Radio Enthusiasts

a. The Role of an 'Elmer': Learn approximately the ham radio way of life of mentoring, known as "Elmering", and why it is critical.

b. Organizing Workshops and Classes: Tips and recommendations to set up academic training for green humans.

c. Sharing Your Story: The importance of community storytelling in inspiring others.

four. Staying Updated with the Evolving World of Ham Radio

a. Subscribing to Magazines and Journals: Recommendations for pinnacle ham radio publications.

b. Joining Online Communities: Dive into forums, on line corporations, and internet websites devoted to modern-day developments inside the ham radio worldwide.

c. Attending Conventions and Hamfests: Experience the bigger ham radio community, discover new era, and meet fellow enthusiasts.

d. Continuous Education: The importance of ongoing education durations, workshops, and webinars.

www.ingramcontent.com/pod-product-compliance
Lightning Source LLC
Chambersburg PA
CBHW071442080526
44587CB00014B/1953